STO

ACPL ITEM
DISCARDED

S0-DSF-882

REPAIR

**DO NOT REMOVE
CARDS FROM POCKET**

ALLEN COUNTY PUBLIC LIBRARY

FORT WAYNE, INDIANA 46802

You may return this book to any agency, branch,
or bookmobile of the Allen County Public Library.

BEGINNER'S GUIDE TO TV REPAIR—2nd Edition

Other TAB books by the author:

No. 648   *Everyman's Guide to Auto Maintenance*

No. 1013
$9.95

BEGINNER'S GUIDE TO TV REPAIR—2nd Edition
BY GEORGE ZWICK

TAB BOOKS
BLUE RIDGE SUMMIT, PA. 17214

Allen County Public Library
Ft. Wayne, Indiana

SECOND EDITION

FIRST PRINTING—SEPTEMBER 1979
SECOND PRINTING—APRIL 1980

Copyright © 1979 by TAB BOOKS

Printed in the United States of America

Reproduction or publication of the content in any manner, without express permission of the publisher, is prohibited. No liability is assumed with respect to the use of the information herein.

Library of Congress Cataloging in Publication Data

Zwick, George, 1910-
  Beginner's guide to TV repair—2nd Edition.

  "TAB # 1013."
  Includes index.
    1. Television—Repairing—Amateurs' manuals.
I. Title.
TK9965.Z87   621.388'8'7   79-17429
ISBN 0-8306-9758-6
ISBN 0-8306-1013-8 pbk.

Preface

To serve a useful purpose a preface should be a sort of abstract or summary stating what a book is about as well as what it is not. For the sake of absolute clarity we shall first state what this book is not. First, it is not a complete course in TV repair. It will not teach you how to do jobs that can be done only by a competent, experienced TV servicer. This includes many and varied tasks, for example, the replacement of the picture tube and the subsequent adjustments—a task definitely not in the beginner category.

Second, this book will not teach or recommend any repair involving personal hazard. There are such repairs in the complex machine called a TV set, particularly a color TV set, and no man, however handy or mechanically inclined, should be foolhardy enough to attempt any of these.

Finally, this book will not suggest any repair in which there is a low probability of success. To do so would likely to more harm than good because it is liable to condition the TV owner to accept a mediocre, substandard level of performance, rationalizing that this is "the best he or she can get."

What does the book intend to accomplish? It is the purpose of this book to enable anyone interested enough to learn the basics of TV repair to keep his or her receiver performing at its best by following a safe, simple, preventive and corrective maintenance program that requires no specialized tools or equipment. This maintenance includes such things as adjustments to compensate for aging, deterioration, image distortion, etc. It also tells how to correct such defects as improper illumination (brightness/contrast), picture instability (rolling/tearing), and overloading (smearing, etc.). In addition this book includes procedures on how to remedy such catastrophic defects as loss of vertical (decrease in picture height to a thin horizontal line), loss of sound, loss of picture, loss of both sound and picture, as well as such frightening symptoms as smoke, crackling and frying sounds, etc.

In the case of color receivers the uninitiated will find that this book solves the mystery of color without creating new technical mysteries by showing that the color TV receiver is basically a black-and-white set with added capabilities.

With the advent of solid-state technology in TV receivers, whether hybrid (part tube, part transistor) or all solid-state type, the do-it-yourself TV owner must of necessity acquire a modicum of familiarity with the infrastructure of this type of equipment. This involves not only the gradual disappearance of vacuum tubes and their replacement by semiconductors—transistors, diodes, varactors, etc.—but also the almost-revolutionary change in construction through miniaturization. Bulky discrete parts with their point-to-point wiring have been almost totally replaced by miniaturized subassemblies, printed-circuit wiring and integrated circuits (ICs). As a logical consequence the handling of such components has also undergone a radical change. The old familiar screwdriver, soldering iron and long-nose pliers have been superseded by semiminiature, almost jeweler-type tools. Their manipulation requires new techniques and considerable circumspection. This does not mean the end of do-it-yourself. In fact, in many instances it has actually become easier. Yet there still are areas best left to the professional repairman, and I identify those in the book. But I also guide the would-be home do-it-yourselfer through the new required procedures in handling transistors, ICs and printed-circuit (PC) boards. A new chapter, *Solid State and How to Treat It*, is a sort of "minimal adequate" course in the new techniques for the do-it-yourself TV owner.

To establish correlation between functional circuits and physical components—a vital step in identifying actual TV parts and subassemblies by the TV owner—I have shown parts of actual set schematics of modern receivers, sectionalized on the diagram in correspondence with physical circuit boards. This follows the very common practice of TV manufacturers in their service manuals and data; thus, when you look at such a partial diagram, you also see the physical PC board or subassembly of that diagram. Following the same logic I have also shown some actual sample procedures for removal of circuit boards from a complete TV set.

As to the *don'ts* for the TV owner I identify those tasks which for reasons of safety—as in case of picture tube replacement—or of technical complexity—this includes such functions as IF and tuner alignment, rarely needed, but extremely complicated—so that you will not go out on a limb so to speak. While the more courageous TV owner might attempt, following manufacturers' instructions and admonitions, to replace the picture tube, he or she would still risk the predicament of a possible "stuck" tube which refuses to slide out of the yoke assembly. In case of tuner or IF adjustments, although not involving the danger of an implosion as in case of the picture tube, I caution the do-it-yourselfer against almost 100% certain failure, unless you sport an elaborate and expensive array of sophisticated test equipment. In such cases, and wherever feasible, I direct the TV owner to an alternative procedure, as in some cases of tuner replacement.

The technical level of this book is necessarily different for different users. While the more experienced reader may find some of the detailed explanations somewhat naive, these same explanations may be nothing short of astounding revelations to the uninitiated; however, the overall level of the subject matter is such that the average reader will be able to progress from page to page without difficulty, so that what may at first glance seem too technical and overwhelming will in due course become quite logical, fairly simple and relatively easy to accomplish.

Lastly, and by no means negligible, this book is intended to perform a needed service for those who do not ever wish to tinker with a TV set but who would very much like to have a general understanding of a TV receiver: how it works and why it fails, without resorting to engineering or mathematics or even technical language. For there is a great advantage to be gained from an intelligent appreciation of how any household appliance works, especially so a television receiver, even if only to be able to talk intelligently to a professional servicer, as well as listen to him or her understandingly.

George Zwick

Contents

1 The TV System 11
Basic Sound & Picutre Systems—Original Sound to Reproduced Sound—Original Picture Image to Radio Signal—TV Sound—Complete TV Signal

2 How a TV Receiver Works 25
Antenna System—Transmission Line—TV Receiver—Solid-State TV Receiver—Tube-Type & Solid-State Receiver Comparison—Integrated Circuit Applications

3 Color TV & How It Works 51
Nature of the Visual Spectrum—Color Transmitter—Color Receiver—Color Picture Tube—Recent Color Tube Improvements

4 Color-Only Sections 67
Burst Amplifier—Color Sync & Oscillator Control—Color Oscillator—Color Demodulators—Color Amplifiers—Color Killer—Convergence—Fixed Static Convergence—Color Picture Tube Circuits

5 Introduction to Troubleshooting 79
Equipment—Tube Pulling—Testing by Substitution

6 Operating Adjustments 95
Vertical Size—Vertical Linearity—Vertical Roll—Freak Sync Troubles—Horizontal Size or Width—Horizontal Linearity—Horizontal Drive—Pincushion Correction—Centering—Focusing—Ion Trap—Magnet Precautions—Other Horizontal Defects—Tuner Spray Lubricant

7 **Internal Troubleshooting** 127
Blooming & Defocusing—Dim & Blooming When Set Is Cold—Hum & Buzz—Multiple Pictures—Line Noise & Line Noise Filters

8 **Troubleshooting Catastrophic Failures** 141
Completely Dead Receiver—Tubes Light Only—Raster Only—No Sound But Picture OK—Snow—Sound OK But Picture Missing—Dim Picture Not Covering Full Screen—Picture Dim & Brightness/Contrast Ineffective—Heavy Black-and-White Bars—Picture Snaking—Sound Bars—Herringbone Weave—Barkhausen Interference—Picture Ghosts—Loss of Fine Detail—Smear—Line Pairing—Missing Adjustments—Short or Fire

9 **Troubleshooting the Color Section** 169
No Color But B&W Normal—Weak Colors—Color Functions—Individual Color Adjustments—Color Interdependence—Red Missing—Blue & Green Missing—Blue Missing—Color Presentation Unstable—Color TV Antennas—Combination VHF-UHF Antennas—Color Smear—Degaussing

10 **Solid-State Circuitry** 193
Effect on TV Design—Service Techniques—PC Wiring—Example & Illustrations

11 **Typical Trouble Symptoms & Their Probable Causes** 201
Set Completely Dead—Raster With No Sound or Video—Sound But No Raster—Raster & Video But No Sound—Loud Arcing or Popping Noise—No Red & Hue Control Has No Effect—Picture Is Reduced Overall—Horizontal Line Only—Raster Reduced Top & Bottom—Raster Appears as a Small Circle—Lack of Width—Too Much Snow in Picture—Horizontal Weave With Color Bars—Movable Ghost When Fine Tuning—Vertical Roll—Smoke or Fire—Poor B&W on Color Receiver—Picture Pulling Horizontally at Top—Barber Poling or Poor Color Sync—Poor Focus in Color Receiver—Misconvergence—Reduction in Sound After Warmup—Improper Tint Control Range—Sound Beat Pattern in Picture—AC Hum Bar Floating in Picture—One Color Missing From Picture

Glossary 207

Index 221

Chapter 1
The TV System

Before embarking on a detailed, nontechnical description of various portions of a modern TV system it is desirable to present a brief synopsis of the overall process, from end to end, in order to provide a continuity of visualization so that as you read about a particular step in the system you will have in the back of your mind, so to speak, the complete picture of the overall purpose and end result to be achieved.

BASIC SOUND & PICTURE SYSTEMS

The TV system, whether monochrome (black and white) or color, consists of two distinct transmitter-receiver systems—the sound (audio) system and the picture (video) system. The interconnections and any commonalities between the two systems is entirely incidental. In other words it is not essential that there be any common connections or functions between the two systems. It is done merely for economy and convenience. Let it be thoroughly understood that a physically separate sound system alongside a corresponding totally independent picture system would be just as feasible from a strictly technical viewpoint.

The sound system is essentially the same as the FM portion found in AM/FM receivers. The FM sound part of the TV signal picked up by your antenna has the same general characteristics of an FM signal—namely, high immunity to noise (static, electrical interference, etc.), relatively short-range reception capability and (po-

Fig. 1-1. Simplified block diagram of a sound transmitter (A) a sound receiver (B). Notice that the two systems are virtually mirror images of each other.

tentially) a higher sound quality. Figure 1-1 shows, in block diagram form, the major components of the TV sound system.

Figure 1-2 shows, in similar form, the building blocks of the video (picture) system. This is an AM signal (similar to the AM music stations) and is subject to the same general conditions accompanying AM station reception—fading, freak long-distance reception, interference (streaks, etc., across the screen) and so on.

Terminology

A few words about the terminology we just used might not be out of order here. AM stands for *amplitude modulation* and signifies a system of radio transmission and reception in which the magnitude of the signal varies directly with the loudness of the voice or music or with the brightness of the image in case of a TV picture. Since electrical noise can add to or subtract from such a signal it is only logical that such noise is reproduced by the receiver as physical noise (static, clicks, etc.) or as streaks, dashes, etc., on the TV screen. In other words AM is not immune to noise.

FM stands for *frequency modulation*. This system ignores any variations in size of the radio signal as might be caused by the addition of noise to the signal and transmits and receives sound information by means of a variation in the *spacings* (modulation) between adjacent waves. Technically, both picture and sound could be AM or FM, but there are other reasons for the present use of FM sound and AM picture.

Transmitting & Receiving Systems

In Figs. 1-1A and 1-2A the vertical dashed lines define the three distinct major functions of each transmitting system. On the extreme left of Fig. 1-1A is the microphone which changes sound waves to electric currents. In Fig. 1-2A the camera performs the corresponding function of changing light into electric energy.

The second sections of Figs. 1-1A and 1-2A serve the major functions of amplification (enlargement) of the faint signals. The third sections in both illustrations convert the amplified signals into a form suitable for transmission via the antennas. They also further amplify them before feeding them to the antennas. The right-hand sections in both illustrations actually radiate the signals into space to be received by many receiving antennas.

Receiving systems (Figs. 1-1B and 1-2B) are virtually the mirror opposite of the transmitting systems just described. On the extreme left are the receiving antennas, located so as to "get in the

Fig. 1-2. Simplified block diagram of a picture transmitter (A) and a picture receiver (B).

way" of the radiated energy from the transmitters. (That is why receiving antenna location is so important.) The second section of each receiver amplifies or builds up the minute signals to the levels necessary for further processing. The third section in each illustration converts the signals back to a form suitable for display as picture or sound, as well as amplifying them. The right-hand sections reproduce sound waves from the electrical energy in the sound signal and form pictures from the electrical energy carried in the picture (video) signal.

ORIGINAL SOUND TO REPRODUCED SOUND

All sound (voice, music or noise) is produced by a physical force which sets the air into vibration. The amount of force applied to the air determines the loudness of the sound, while the different pitch (from bass to a whistle) results from the rate (speed) of vibration of the air.

Microphone

The microphone is a device which converts mechanical energy (air pressure or vibration) into electrical energy. The amount of the electrical energy produced depends on the physical force applied (sound loudness). The pitch of the original sound determines the frequency (numerical rate of vibration) of the electrical energy produced. For sound the frequency may be as low as 20 vibrations per second (a deep, organ note) and as high as 20,000 vibrations a second (a very high-pitched whistle). Incidentally, ordinary household AC has a *frequency* of 60 vibrations (cycles or hertz) per second.

Amplifier

These electrical devices, using either tubes or transistors, have but one function—to faithfully enlarge the feeble electrical currents emanating from the microphone. No one tube or transistor can provide all the amplification that is required; so there usually are a number of amplifiers, following each other, each amplifying the output of the preceding amplifier or stage until the required maximum is obtained.

Audio & Radio Frequencies

To attain a useful understanding of TV transmission and reception it is essential to have a clear concept of frequency. As was mentioned earlier electrical energy exists as a wave-like phenomenon, just as sound energy consists of air waves, and not unlike the waves or ripples caused by a disturbance in water. The term frequency is used to denote the number of times a wave recurs in a certain period of time, usually 1 second; thus, in all future reference to frequency in this book the meaning will be the same: the number of waves per second, or, preferably, cycles per second. As stated before, audio frequencies (AF, sound frequencies audible to the ear) are assumed to extend from about 20 cycles per second (deep organ note) to about 20,000 cycles per second for a very shrill whistle. Above 20,000 or 30,000 cycles, they are called radio frequencies (RF for short), extending through AM and FM radio, TV and radar transmitting and receiving frequencies to the near-optical (light) portion of the spectrum. Here we find first infrared (invisible heat waves), followed by the visible light spectrum (red through violet), the ultraviolet (again invisible) and on to X-rays, etc. Video (picture) frequencies are related to television. They stand for those frequencies which carry picture information and extend from about 20 cycles per second to about 4,000,000 cycles per second.

Until several years ago frequencies in the entire spectrum from audio through invisible light we referred to as cycles per second, thousands of cycles per second (kilocycle), millions of cycles per second (megacycle), etc. Now by international standards the term hertz, abbreviated Hz, is used instead of cycles per second. Heirich Hertz, a German physicist, was the first to demonstrate the production and reception of electromagnetic or radio waves. Thousands of cycles is referred to as kilohertz (kHz), etc.

A final note on waves and frequencies. The higher the frequencies, the shorter the length of the wave. For example, the sound wave of a telephone bell (with a frequency of, say, 1000 hertz) is approximately 186 miles long. By contrast the radio wave transmitted by a UHF television station is little more than 1 foot long. This explains why the elements or metal rods of a UHF TV antenna are so much shorter than the corresponding VHF types. While any frequency can be transmitted from place to place over wires, only those classed as radio frequencies (RF) can be sent through space (wireless transmission).

RF & AF Amplifiers

The middle sections of our simple diagrams in Figs. 1-1 and 1-2 consist mainly of amplifiers, accessory devices and controls. Those immediately following the microphone are audio amplifiers, as stated previously; however, in order to transmit this audio economically and efficiently over great distances a temporary change is made. The audio wave is superimposed (that is, it modulates) on a radio wave piggyback style. This combined RF+AF energy is further amplified and finally applied to the antenna.

Antennas

A transmitting antenna has but one function: to radiate the energy applied to it, in the desired direction (or in all directions), in the most efficient manner possible. While the obvious purposes of the receiving and transmitting antennas seem to be quite different from each other they actually behave very much alike. Incidentally, despite many claims to the contrary, there is no substitute or shortcut worthy of its name for a high-grade, elaborate antenna, whether for transmitting or receiving.

At the receiving end the antenna does not radiate energy—it collects. Located in the path of the radiated energy the receiving antenna collects or acquires a very small sampling of the original energy or signal as radio waves cross it. By means of a transmission line (sometimes called lead-in) the intercepted signal is carried to the input of the receiver.

It is unfortunate but true that while a properly designed receiving antenna will favor the stations (frequencies) for which it was designed and discriminate against others, it is fairly helpless in rejecting undesirable electrical energy, such as noise, static, etc. The transmission line, or lead-in, is equally susceptible to noise or static pickup, although something can be done here. Shielding, or protecting the lead-in from surrounding electrical noise, is feasible. The antenna itself cannot be shielded (or it won't receive any signals); the only recourse lies in the selection of a suitable location where noise is at a minimum.

Receiver

The amplifier sequence in the receiver is virtually a mirror reflection of that in the transmitter. Amplifiers immediately following the antenna amplify the combined (RF+AF or modulated RF) signal to the required level. Next, and in the same general portion of the receiver, a reversal of the modulation process takes place by a device called a demodulator or detector. This reversal consists of stripping the RF, which has served its purpose, from the signal, leaving only the audio-frequency signal. Further amplification now takes place until the AF signal is sufficiently strong to operate a speaker or headphones.

Speaker

In function, a speaker or loudspeaker is the exact opposite of the microphone. Electrical energy in the form of audio waves is applied to the speaker, where it is converted to the mechanical motion of a surface usually called the speaker cone. The cone vibrates faster or slower, depending on the frequency (pitch) of the corresponding original sound waves, and sets the air in similar motion. The loudness of the sound depends on the distance the cone moves, which in turn depends on the strength or size of the electrical wave applied to it.

ORIGINAL PICTURE IMAGE TO RADIO SIGNAL

In the following paragraphs I shall trace the sequence of steps necessary to convert a visual scene into an electromagnetic energy wave suitable for transmission in space. After the conversion from light to electrical energy has been accomplished in the camera the remaining functions are very similar to the sound-to-RF-to-sound process just described. The functions to be described apply equally to monochrome and color transmission and reception and should give you a clear nontechnical concept of what is going on so that you will be able to deduce what went wrong and where in the event your TV set performs unsatisfactorily or fails to perform altogether.

Physical Image

All images—persons, scenes or whatever—are seen by the human eye because of the light they emit or reflect. It is quite obvious that all such images are color images. In addition the various colors reaching the eye are of different light intensity; thus, we commonly say dark brown, or dazzling white, or bright yellow, etc., suggesting that we see, in addition to different colors, different degrees of light intensity. The best example of this is a so-called black-and-white picture made by a camera, where all colors become different shades of gray. A person's face is brighter than his or her shoes, but not as bright as a white shirt, etc. In a manner of speaking our descriptions of different colors are given in terms of brightness or dimness, with the notion of color being a separate characteristic of the object or scene we are viewing.

Photographic Recording

To the human eye a scene or object presents a continuous range of color and illumination. Actually, however, this seeming continuous image is but a composite of many discrete little areas, each of which may differ from its neighbor either in color, in brightness or both. In a black-and-white photograph the image is actually made up of tiny specks of the image spaced from each other. The concept of grain (fine-grain photograph, for instance) refers to just this particle structure. While the color photograph is claimed to be grainless, this is but a relative term in contrast with the chemical grain structure of the black-and-white picture.

Halftone

A practical example of a discrete particles composition of an image is the familiar newspaper or magazine photograph. Viewed with the naked eye such a photograph looks smooth and continuous. Under a magnifying glass, however, the picture is revealed as consisting of individual dots with clear spaces between dots. Shading in the picture depends on the weight of the black dots in comparison to the adjacent white spaces. While this composition is due to the technique of mass reproduction (newspaper printing, etc.), the fact remains the images of very high quality can be and are reproduced by the discrete dot structure.

Conversion of Image Dots Into Electric Signals

The first step in producing a television signal is the conversion of light (optical energy) into electricity. It begins with a photographic-type camera, lenses, focusing, viewfinders and all; however, here is where the similarity ends. Instead of projecting the image onto a ground glass (as in studio cameras) or on a photographic film (any camera) it is projected onto a special electrochemical surface called a photomosaic. Incidentally, the projected image may be quite small, often not much larger than a postage stamp.

Next an electron beam or stream scans the mosaic image, one spot at a time, as if reading the bits of the image. Each time the electron stream impinges on one of the spots of the mosaic image an electric current is generated having characteristics peculiar to that particle; that is, the electric pulse or current from that image particle contains information necessary to reproduce this image particle on the home TV screen.

There is also included with each such electric pulse the position or location information, both in the vertical and the horizontal directions. This positioning information is transmitted and received by the home TV antenna, faithfully amplified and reproduced and finally used to guide the electron beam in the TV picture tube for proper positioning of the image particle.

A simple example of an electron gun used to produce a scanning beam is found in the neck portion of a TV picture tube (see Fig. 3-6). It is composed of a glowing heater—a sort of electron generator—and a number of metallic structures (hollow cylinders, baffles, etc.) which shape and guide the electrons toward the screen in the form of an extremely thin, sharp beam. In the home receiver the electron beam "writes" on the picture tube screen in step with the "reading" done by the beam in the TV camera. The location and brightness of each written bit is automatically known from information transmitted with each bit so that the completed written image is an exact duplicate of the original. The scanning sequence is arranged in accordance with an established code (National Television Standards Committee, NTSC). In simplified form it is as illustrated in Fig. 1-3 and may be described as follows. (This scanning sequence may be

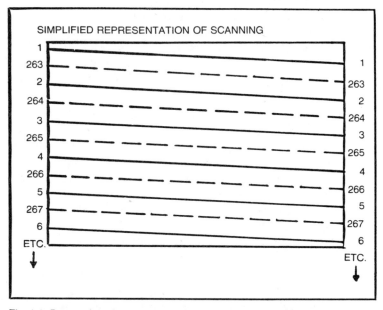

Fig. 1-3. Drawing showing how a TV picture is scanned by electron beams. The solid lines represent one scan field, the dashed lines the alternate scan field. Line spacing and slant are exaggerated for clarity. Dashes are used to distinguish between the two fields. The sequence is 1, 2, 3, 4, 5, 6, 7, 8 and so on, up to about 262. Then it continues with 263 and on to about 525.

considered, for the sake of illustration, analogous to the writing of a letter, double-spaced, then filling in the spaces between the lines so that the complete letter is read line after line, from top to bottom):

Beginning at the upper-left corner the electron beam scans line 1 from left to right. It then skips back to the left side and scans the second line underneath the first. This is line 2. This sequence continues until approximately 235 lines have been produced. In Fig. 1-3 all these lines are shown solid, for a reason which will become clear presently. The beam continues to scan an additional 27½ lines, for a total of 262½, with but one difference—these additional lines are not visible on the home screen—their location is just below the visible bottom of the picture. We shall later explain the purpose of these "invisible" lines. The beam now skips back to the upper-left corner and starts scanning again, but this time between the lines just finished, producing lines 263, 264, etc., until about line 498, then continues scanning invisible lines again until 525 are produced. The complete process now repeats, startling with line 1.

The picture tube screen has now been scanned twice, each scan lasting 1/60 of a second, the period of which is called a *field*. The two fields, lasting a total of 1/30 of a second, are called a *frame*. The process of alternate line/field scanning is called *interlacing*. In case of a still scene, more than one frame of the same scene would be scanned; in action scenes, there will always be at least one frame (two fields) scanned, this being required for a correct picture of the scene.

Blanking & Synchronizing

In describing the scanning sequence it was stated that at the end of each line, as well as at the end of each field, the beam skips back to the beginning of the next line, and, at the end of a field, to the upper-left corner for the beginning of the next field. I shall now explain *blanking* as well as the very important activity during the blanked interval. Blanking simply means making the lines invisible since they are not part of the image and would only add meaningless lines, bars and streaks to the picture. This important activity consists of transmission of *synchronizing* pulses, guidelines for starting and ending of each line. These are sent at the end of each line, while the timing of each field is sent during the blank period between fields. Any deficiency or degradation of these *sync* pulses would be cause for picture instability, such as horizontal tearing or vertical rolling. There are other signals sent during the blanking period, such as color correction (VIR) in the most recent expensive TV sets. I shall refer to this later in the book.

In connection with this discussion of blanking it should be mentioned that in addition to the brightness and contrast controls on the home TV set, which are adjustable at will, there are some definite normal levels of brightness and contrast, preset and identified by the transmitted signal, incident to the blanking interval information; thus, it is possible for the home TV viewer, by misadjusting the brightness control, while at the same time misadjusting the vertical hold control so that the picture begins to roll slightly, to actually see the vertical blanking and synchronizing pulses on the screen, located between adjacent picture frames. Of course, such excessive brightness setting is detrimental to normal picture light distribution, causing light areas to be burned out with loss of detail. It similarly affects the opposite end of the illumination range, preventing black areas from being actually black. This applies equally to color as to black-and-white sets; thus, indirectly, the appearance of any such blanking lines on the screen is a sign of excessive brightness in some TV sets, assuming an otherwise normal set.

Interlace & Pairing

It was mentioned previously that the alternations of lines from the first and second fields of each frame is called interlacing. Later in the book I explain the related phenomenon of *pairing*, an operational defect which causes picture degradation. It is therefore desirable to explain this relation a bit further at this time. I shall do so with the aid of an analogy to the familiar photograph.

If an ordinary photograph were to be cut into a large number of very thin strips the original image could still be seen, no worse than before, provided the individual strips were kept from moving, sliding or shifting. If, however, some or all of the strips were moved so that they partially or completely overlap the picture would naturally suffer, depending on the degree of overlap. With regard to the TV picture it is quite correct to visualize it as being made up of such very narrow strips (called lines) arranged in two steps. In step one every other strip is temporarily omitted. This produces a picture with a number of gaps. In step two these gaps are filled with the remaining set of strips, each exactly between the two adjacent strips. The picture is now complete.

If, however, the second set of strips were placed so that they partially or even totally overlap the first set the picture will not only be incomplete because some of the strips are covered over, but also because there is a mixing or blurring of whatever is still visible. This phenomenon of overlapping and blurring is called pairing since there no longer are any single, individual lines, but pairs of such lines. As explained later such pairing is caused by a failure of the horizontal sync, either due to a defect in the TV set or because of transmission difficulties, making each line double-width and resulting in but half of the normal number on the screen.

Automatic Brightness Control

In recent years manufacturers have added a number of so-called automatic features to their receivers, usually for the sake of simplification of operation by the user. Sometimes, however, such a feature is a protective measure against improper adjustment and possible consequent damage to the set. Such is the case with ABC (automatic brightness control) or ABL (automatic brightness limiter). From a technical viewpoint this feature protects some internal parts, especially the picture tube, from premature failure from excessive cathode current; it also keeps the image on the screen from becoming extremely overbright due to careless setting of the front-panel brightness control. As a consequence the above mentioned trick to display the blanking and sync bar may not work without a bit of trying and careful misadjustment. It is assumed that after such a trick the set owner will return the controls to normal by first reducing them to below normal, then gradually bringing them up to normal.

Conversion of Electrical Impulses into Radio Waves

Electrical impulses representing individual image elements and timing (synchronizing) pulses are combined with or superimposed upon a radio wave (carrier), amplified to the required power level required for transmission over the intended distances (a TV station is licensed for a certain maximum power level to meet the needs of the geographic area it is to serve) and applied to the antenna.

Incidentally, frequencies of TV stations in the United States are much higher than those used for ordinary AM radio broadcasts for technical reasons (number of stations to be accommodated, shorter distances covered, etc.). While ordinary AM stations are centered around 1 MHz (megahertz), TV frequencies range from just over 50 MHz to almost 900 MHz. This affects not only the location of TV stations in the radio spectrum but also poses different requirements on design, complexity, etc.

TV SOUND

There remains but one additional signal required for a complete TV transmission and that is the sound (or noise) accompanying the picture. As stated before TV sound is transmitted as FM for certain technical reasons as well as for some advantages (noise immunity, for instance). TV sound could be transmitted over a completely separate system from microphone to loudspeaker. This, however, would be unwieldy, more expensive and technically less satisfactory. For these and other reasons TV sound is sent as part of the overall station transmission with many functions, circuits and actual components shared by the picture and the sound portions of the receiver.

Figure 1-4 shows, in simplified form, the sequence of functions and their combination into a single final signal for transmission. Notice that the sound part of the system resembles very much the

Fig. 1-4. Simplified block diagram of a TV transmitter, showing separate sound and picture paths combined before reaching the transmitting antenna.

23

ordinary radio (AM) mentioned previously; however, instead of being applied to the antenna system directly, the sound signal (an FM system) is combined with the TV picture signal; then both are amplified as an integrated signal and applied to the transmitting antenna. For all practical purposes this is now a single radio signal containing all the information necessary to reproduce a TV picture, both sight and sound.

COMPLETE TV SIGNAL

To summarize the characteristics of a typical TV broadcast signal as received by the home TV set, the TV signal is a composite of two independent signals or stations—an FM sound station and an AM picture station. The FM sound station contains just the sound that accompanies the TV picture. The AM picture signal consists of the following:

1. Picture brightness information, bit by bit.
2. Picture background illumination, the overall scene.
3. Picture bit location information in the form of cueing signals for starting and stopping each line, field and frame.
4. Picture repetition rate timing signals which insure exactly the same line, field and frame frequency at both the transmitter and receiver.

Items 3 and 4 above are, in effect, two components of control information serving the same general purpose; that is, to insure that the reproduction of the picture on the home TV set is precisely in step with the production of the original picture in the pickup camera. During the description of a TV receiver in Chapter 2 reference is made to the various components of the timing signals, particularly in connection with the subject of *vertical hold* and *horizontal hold*.

Chapter 2
How a TV Receiver Works

If you were to look at a wiring diagram, called a schematic of a TV, it would seem to be a mass of meaningless lines and symbols, but to an experienced TV servicer every line and every symbol is significant. They help the servicer follow the path of each signal through each section of the receiver in order to determine why the set won't work as it should. A schematic also tells the technician what should happen to the signals as they pass through each section. To help you get some idea what's going on in those circuits let's consider each part of the TV receiving system.

ANTENNA SYSTEM

A home TV antenna differs from that used by a TV station for two major reasons. One is the fact that the receiving antenna must be receptive to a large number of stations, each of a different frequency. In other words the receiving antenna is a broadband device; therefore, some compromise is necessary in comparison to an antenna designed for one particular station only. The latter, of course, is the case with the transmitting antenna of any one station. The second difference stems from the fact that the transmitting antenna is intended to radiate in all directions—it is omnidirectional—while the home antenna, even if it has to be aimed in different directions to receive stations from different geographic locations, is in fact a highly directional system. Its multidirectional receiving capability is provided by rotating it to the desired position.

It may be said that all the elements (rods) of the antenna point in the same direction for maximum signal pickup from that direction.

TRANSMISSION LINE

A brief comment on the function of the lead-in or transmission line. Its sole purpose is to transmit the energy picked up by the antenna to the TV set input, with a minimum of loss and, perhaps just as important, with no pickup along the way. Since any such pickup is likely to be almost all noise and hardly any signal the importance of a proper high-quality transmission line cannot be overemphasized. We shall see later, in troubleshooting, how a poor transmission line, due to poor quality, installation and location, can be the cause of inferior TV reception.

TV RECEIVER

Figure 2-1 is a block diagram of a TV with each of the major functions of the receiver represented by a box. We shall now examine each of these boxes primarily from the service viewpoint.

Tube-Type Tuners

Boxes A1 and A2 represent devices called tuners or *front ends*. Their purpose is to select the desired station and amplify the relatively feeble signal from the antenna. For the past few years there are two tuners in each TV set, one VHF (very-high frequency), covering channels 2 through 13, and UHF (ultra-high frequency), covering channels 14 through 83. Both tuners are very complex, rather sophisticated assemblies, requiring for their maintenance and repair test equipment and technical know-how well beyond the capability of not only the beginner but even a good segment of the TV repair industry. In fact, many high-grade TV repair shops have TV tuners repaired by special service stations fully equipped to do the work.

The above statements are intended to caution the beginner against any rash action in attempting to correct a malfunction in this portion of the receiver. There are, however, certain tasks which are within the ability of the beginner that can be performed satisfactorily, and these are described in chapters covering troubleshooting.

Construction and location of the tuner are of particular interest, more so than any other section of the set, except perhaps the high-voltage cage, which is described later. Both the station selector and the fine-tuning adjustment are, for technical reasons, physically

Fig. 2-1. Block diagram of a typical TV receiver. The lines and arrows indicate signal paths.

built into the tuners. This requires that the tuner be located at an accessible position on the cabinet. For this reason, as well as for some strictly technical factors, the tuners are separate physical entities, little box-chassis which are relatively easy to disconnect from the remainder of the TV set.

Interconnections between the tuners and the main TV chassis and the antenna are as follows:

1. The signal input from the antenna terminals to the tuners is by way of twin lead or sometimes concentric (coaxial) cable.
2. The signal output, almost universally in the form of a shielded coaxial cable (a center wire surrounded by insulation with a braided copper outer sleeve) terminated in a single contact plug.
3. A three- or four-wire cable connecting to the various voltages in the main chassis. These wires may end in individual lugs fastened by screws, or in a common, multipin plug which fits a mating socket on the main chassis.

Of course, there is some provision for rigidly fastening the complete tuner assembly to the cabinet. Figure 2-2 is one example of such a tuner; others may differ in a minor detail, but essentially they are of the same general construction and interconnection arrangement.

The VHF tuner, if it is of the tube type, usually has two tubes of the miniature (all glass, no separate base) kind, and they are clearly marked for location, type and function; thus, a tuner may have the following stamping near the tube sockets: 6AH5, V1; 6EA8, V2. Sometimes, when more than one tube type is suitable, dual markings may appear, such as 3HA5 or 3HM5 (3HA5/3HM5) and 6EA8. In other cases the designations V1 and V2 may not appear on the tuner, but will be shown on the tube location chart somewhere inside the TV cabinet. The correlation between tube markings on the tuner and the tube location chart inside the TV cabinet is used extensively to identify tubes by function, malfunctions, symptoms and suggested corrective action. I shall have more to say about tube locations and charts under troubleshooting procedures.

In those tuners where transistors have replaced tubes, there is the high probability that the transistors are soldered into place instead of plugged into a socket, in which case the advice given above regarding tuners in general applies very much here. To repeat, tuner work, other than tube replacement, is not for the amateur. Even if a beginner could locate the suspected transistor, any attempt

Fig. 2-2. Typical VHF tube-type TV tuner, showing input connections (1), coaxial output connection (2), and a tube shield (3).

to unsolder it, however carefully done, is sure to be disastrous. It takes an experienced hand and special tools to successfully solder or unsolder semiconductors. Removal of the entire tuner, where such a procedure is indicated, if not done carelessly, is far easier and safer.

Of the two tubes in a tube-type tuner, at least one is a dual-purpose type; so there are three distinct functions performed here. The first tube, almost always located nearer the incoming antenna lead-in, is an RF amplifier which increases the signal level of the station being received. A bad tube in this position will seldom result in no picture and sound whatsoever. Instead, it will account for a very feeble picture with heavy snow and weak sound very much resembling a weak or distant station. The remainder of the TV set is doing its best with the very small input signal that manages to get through the defective amplifier. Removing a bad RF amplifier tube from its socket may produce very little added degradation of the picture. In those sets where removing one tube causes all other tubes to stop glowing, another procedure is suggested under a discussion of transformerless TV receivers.

The second tube of the tuner is almost always (except in those rare cases where the tuner has three tubes) a dual-function tube. It will invariably be identified on the location chart as V2, and sometimes with the added optional description, such as MIX-OSC or mixer-oscillator. We need not know the whole story about mixer-oscillators to know their functions. If this tube fails there is nothing on the screen (except the white *raster*), no faint picture, no snow, just a total blank, and no sound. On a nontechnical basis one might consider a defective V1 (RF amplifier) as a gate that is barely cracked open, while a defective mixer-oscillator is shut tight (100% closed) so that nothing can get through.

I mentioned the rare possibility of encountering an older vintage TV set having a three-tube tuner. These tubes probably will be identified on the chart as V1-RF, V2-MIX, V3-OSC. It should be apparent from this that the combined functions of V2 in the modern tuner are separated in the old three-tube unit. The comparison is quite simple. The first tube is identical, both in function and malfunction, on both tuners. The tube marked V2-MIX in the old version, when defective, may still permit some noise (snow) to get through, but seldom a picture, even a faint one, or sound. A failure in V3-OSC may produce similar symptoms with one exception: no picture or sound at all. In troubleshooting it is not automatically necessary to replace both mixer and oscillator tubes if only one tests unsatisfactorily, but it is most practical to suspect both tubes when the symptoms are as described above.

The Semiconductor Tuner

In recent years the change from tube to solid-state TV receivers was made in two steps. The first was to the hybrid (tubes and semiconductors) receiver, the second to all semiconductor set. Strictly speaking, there are no 100% solid-state receivers since at least the picture tube, and often one or two others, are still of the glass type. But these sets are commonly described as all solid state. While a few hybrid sets are still being manufactured the tuners in virtually all TV sets are of the solid-state variety. The reasons are many and varied, but as far as the nontechnical consumer is concerned this feature is all the better. These tuners are more stable, more rugged and longer lasting. Figure 2-3 is a view of a modern solid-state tuner. Greater compactness and freedom from protruding tubes is obvious even if the durability and greater freedom from repairs may not be visually evident. I shall get back to this type of tuner when comparing tube-type sets and solid-state receivers.

Varactor Tuner

A radical departure from conventional tuners, whether tube type or all solid state, is the so called varactor type, or voltage-controlled type. It has many advantages over the conventional type—in stability, in freedom from noise, in minimal degradation with use, not to mention in greater simplicity in construction and maintenance. For an understanding of this, however, it is necessary to have a little background information of tuned circuits in general, and TV tuners in particular. A brief such explanation follows.

The term tuning, whether applied to automotive adjustment or musical instrument adjustments, refers to setting or adjusting a device or a group of devices to a predetermined condition. In electronics tuning is used exclusively to indicate adjustment of one or two components in a radio receiver (station tuning) or TV set (channel selection). In repair procedures a function corresponding to tuning is called *alignment*. In each of these examples tuning is accomplished by a coil (a loop or spool of wire) and a capacitor. In the case of TV tuners, each channel has its own little coil (actually a number of such coils) cut to size and connected to the circuit by means of a multiple contact switch actuated by turning the channel selector. Since the exact configuration of such a coil or coils is essential for repeatable accurate tuning (getting the correct station reception every time) these coils are made as sturdy and as rigid as possible. Nevertheless, long use, shock, vibration and switching cause a gradual deterioration of the whole tuning system, eventually requiring replacement of the tuner, or rebuilding—a tedious and expensive job at best. But the weakest component is the switch, operated by the channel selector knob. The switch contacts, being part of the functioning coil-capacitor assembly, cause the worst

Fig. 2-3. Modern solid-state tuner by RCA. Note the compactness. Longer life can be expected since no tubes are used, which generate heat and shorten the life of components.

degradation with use due to contact corrosion, accumulation of dirt and oily film, etc. To the user this evidences itself in the form of a *noisy tuner*—erractic pictures, frequent need for jiggling the knob to restore the picture, and ultimately complete loss of stability—the station will not stay on channel. The varactor tuner goes a long way toward elimination of most of these problems.

In simple terms the varactor is a semiconductor, a diode in fact, which not only acts as a capacitor, hence a tuning device, but one whose capacitance can be changed over a considerable range through the application of a voltage. The device is sometimes referred to as Varicap (trademark), meaning a variable capacitor. The advantages in TV tuners are obvious. Although station switching by means of the channel selector is still required, switching can now be done in less critical points, making for much greater stability. Simplification of tuner design and construction is a further advantage of this new technique in semiconductor application to modern TV receivers. While the advantage may not be obvious to the average user, especially in a new or fairly new receiver, it will become apparent later, as evidenced by the absence of chronic misbehavior of the tuner, as is the case with the coil-switching type.

IF Amplifiers

We know that the RF amplifier is the first tube of the tuner. It and the amplifiers to be described here, have similar functions—to amplify a small TV signal to a larger one. The difference between the two types of amplifiers is in the degree and level of amplification. The term IF means intermediate frequency and simply refers to amplifiers after the tuner (block B, Fig. 2-1). When IF tubes malfunction, the severity of the snow resulting is much less, but they still behave generally as gates which are usually open and more or less shut during a malfunction.

There are, however, some noteworthy differences from the beginner's viewpoint. First, because the signal levels are so much greater here, the chances of a total absence of the picture and sound are somewhat less; a dim or thin picture will get through more easily. Second, because of the use made of IF amplifiers (some common to both sound and picture, others are strictly for sound) a failure in one IF amplifier may cause a loss of sound without affecting the picture. Thirdly, the physical configuration is quite different. While the RF amplifier is almost universally located on a separate, integral unit (the tuner), the IF amplifiers very generally are located in some sort of string fashion on the main chassis of the TV set.

In Fig. 2-1 the paths traveled by the picture and sound from the antenna to the picture tube and speaker are shown by lines and arrows. It can be seen that for VHF (channels 2 through 13) the signal path is from the antenna, via the VHF tuner to the IF amplifiers. For UHF (channels 14 through 83) the path is from the antenna through the UHF tuner, through the VHF tuner, then to the IF stages. This difference is of importance in troubleshooting, as we shall see later, because the RF amplifier on the VHF tuner is no longer the first stage in the path of the signal from the antenna.

As was mentioned previously a TV receiver is really a two-in-one device, combining a picture (AM) receiver with a sound (FM) receiver. As such many components are common to both. These shall now be identified and described. Block B in audio Fig. 2-1 is a multi-amplifier system for the combined video and audio signals. Although these signals are of comparatively greater strength than those entering the front end (A1 or A2) a failure in any stage of this amplifying system can be just as catastrophic for either picture or sound.

There are usually three separate amplifiers or stages in sequence, as links in a chain. A defect or total failure in any one stage breaks the chain and prevents the signal from continuing on its way. There is usually some feedthrough of signal even when one stage goes dead, as in a case of tube or transistor burnout or other failure, so that some small portion of the normal signal is transferred to the next stage. One symptom of such a failure might be a very weak picture on any station; although some stations may come in better than others largely because of the original difference in signal strength between stations.

Symptoms of snow mentioned in regard to the front end are much less apparent here and may even be completely absent. Another result of such a failure might be picture instability, such as tearing of the picture in a generally horizontal (actually diagonal) direction and possibly a rolling picture (vertical direction). This is due to the fact that the amplifiers in block B also must pass the synchronizing pulses. A failure in these amplifiers will invariably degrade the quality of these pulses to a level below that required for picture holding. In the section on troubleshooting, I shall present concrete procedures designed to help locate trouble in the IF portion of the set.

At this point it is imperative to clearly indicate what the beginner may not and should not attempt. I refer to the IF transformer adjustments. You may see around a half-dozen either rectangular metal cans or uncased spool-like coils or transformers, some of them with a hole or slot that seems to invite a screwdriver blade for turning. These are in fact adjustments, but they cannot be properly done by eye or ear. Required is some very sophisticated equipment and very specific knowledge and expertise, different for each and every adjustment in all TV sets. Worse yet, unlike the adjustment of some circuits which I will recommend later in this book and which can be either repeated or restored to their original position, the adjustment of any of the IF coils is, from the very beginning, a point-of-no return case. Not only is it virtually impossible to know what the adjustment is accomplishing, but it is equally impossible to go back to the starting point.

To further illustrate the futility of such adjustments the professional servicer uses complex equipment which gives a visual presentation of the operation of all the circuits at the same time. Only on such a visual display can an expert servicer observe the effect of each adjustment on the overall picture. Looking at the front of the TV set while twiddling one of the adjustments is almost certain to destroy the normal quality of the picture without showing any immediate change as the twiddling is made. Finally, it is a fortunate fact that these adjustments, except for some very minor effects, are virtually permanent and do not require any correction, except by the professional servicer after replacing a transformer in case of a burnout or other catastrophic failure. To sum up, the best advice is leave the IF adjustments alone. They are probably as they should be.

Video Detector

In the video detector stage (block C, Fig. 2-1), although relatively simple and consisting of but one diode (or one tube) in most cases, a number of functions occur. A semiconductor diode (D2) converts the video IF signal to a form suitable for application to the picture tube. Also, the combined picture and sound signals coming from the common IF amplifiers—V3, V4 and V5—are separated. A failure (rather rare) of this diode stops both picture and sound signals. In some receivers the sound is "picked off" before it reaches the video detector; so a failure of the diode in such sets would not affect the sound. We shall discuss this again later under the sections covering troubleshooting.

Video Amplifier

The video amplifier tube (seldom more than one) in (block D V6, Fig. 2-1) is a video (picture signal) amplifier, building up the picture signal to the level required for application to the picture tube. In addition this tube also amplifies the sound IF signal block E in many receivers and the various picture control (sync) pulses block H. Should this tube fail all three signal components would be affected, as in the case of the defective diode in block C. In case of tube deterioration due to old age, etc., the effects may not be catastrophic; that is, the picture and sound may still be there but on degraded levels. In such a case the picture control (sync) pulses may be inadequate to keep the image from rolling and or tearing. In normal operation the picture signal goes from V6 to the picture tube, as shown in Fig. 2-1.

Sound IF System & Detector

The sound IF amplifier system (block E, Fig. 2-1), including V7 and V8, is functionally very similar to the IF amplifiers in block B, except that only sound signals (FM) exist in this unit. Sometimes the amplifier consists of but one stage (one tube), depending on the design of the particular manufacturer. Another similarity to the IF amplifier in block B is the fact that the sound IF amplifier is also followed by a detector—an FM sound detector, often called a discriminator. The discriminator (block F, Fig. 2-1) may be either a tube or a pair of semiconductor diodes, depending on the particular make of the TV set, and a transformer or coil. Failure of the discriminator tube (or diode) will, of course, cause a complete loss or reduction of sound.

Audio Amplifier & Speaker

Beginning at the discriminator output and all the way to the speaker the signal is known as audio for it consists only of audio frequencies. Where the discriminator consists of a pair of semiconductors, the audio amplifier in block G (Fig. 2-1) may consist of two tubes. In some sets a multifunction tube is used as a discriminator audio amplifier combination, followed by only a single tube called the final (output) audio amplifier in block G. In either case the final audio amplifier connects (via an output transformer) to the speaker.

Failure of either or both of the audio amplifiers will cause a total loss of sound, although sometimes a defective first audio tube (V9 in our case) may allow a faint sound to go through. In such a case the volume control which regulates the amount of signal going into this tube (regardless where the control is physically located) may have some minor effect on the already feeble sound.

Sync Circuits

The sync circuits (block H, Fig. 2-1) contain two or three tubes, some of them being sections of dual types. The purpose of the sync section is to provide the proper timing and placement of the image bits on the screen. As stated earlier this synchronizing information is transmitted by the TV station as part of the picture signal. There are two main types of pulses here: those responsible for the proper starting time of each line on the left side of the screen (horizontal sync pulses) and those controlling the exact start of the picture at the upper-left corner of the tube (vertical sync pulses).

Thus, the first function of the sync circuits is that of a sync amplifier; that is, to increase the sync signal to its required level. There is also a sync separator stage which separates the vertical sync pulses from the horizontal pulses. And there is, additionally, a third function called sync clipping, which, in simplified terms, means the extraction of the sync information from the combined picture-sync signal. For practical reasons, such as economy, most of the sync tubes are dual tubes. An additional function, not necessarily pertaining to sync, is performed by such a "spare" half-tube. Frequently the spare is utilized as a vertical oscillator, to be discussed later.

The sync circuits and tubes have no effect on sound at all, and only indirectly on the picture. Actually the picture information is completely independent of sync performance; however, since the sync signal controls the sequence and location of each picture element, any malfunction in the sync section is certain to produce chaos in the picture, either an up or down rolling of the picture or some tearing in the horizontal direction. It is, therefore, a safe general procedure, whenever the picture will not stand still, to suspect one or more of the sync tubes.

Vertical Oscillator/Amplifier

We know that the picture on the TV screen is "painted" by an electron beam moving in zig-zag fashion, one line at a time, from left to right. It is apparent from this action that as the beam sweeps from left to right it is also pulled downward at a constant rate. This downward movement of the beam is accomplished by a system known as the vertical sweep, consisting of a vertical oscillator and a vertical amplifier (tubes V13 and V14 in block J, Fig. 2-1). The two tubes in the vertical system follow each other in sequence so that either one, if defective, will cause a failure in the downward pull of

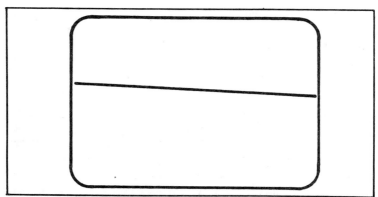

Fig. 2-4. A single horizontal line across the screen is an indication that the vertical deflection system has failed.

the beam on the screen. The first tube, V13, is called the vertical oscillator.

In nontechnical terms the vertical oscillator could be called a timer because it determines how long it takes the beam to traverse the screen from top to bottom. Since the time must be precisely the same as the corresponding time at the transmitter the vertical sync pulse is used to cue the timer. In other words the vertical sync pulse from the transmitter sets the precise rate for the vertical oscillator in the receiver.

Figure 2-4 shows what happens when the vertical sweep circuit fails. With a single bright line across the center of the screen the electron beam is painting all the lines on top of one another. There are other picture defects attributable to trouble in the vertical sweep system. A crowded (compressed or squashed) picture at the top and a stretched picture at the bottom are often due to a defect (nonlinearity) in the vertical amplifier and its associated components.

There are three adjustments associated with the vertical sweep system which may require resetting in case of malfunction. These are vertical hold, vertical size and vertical linearity. The first of these in most sets is accessible from the front of the cabinet. The last two almost always are located in the rear because they seldom require adjustment. Each is described under troubleshooting, where cause-and-effect relationships are outlined.

As the name implies the vertical hold is used to stop this picture from moving up or down; that is, to prevent rolling vertically. The vertical size is an adjustment to make the picture cover the full height of the visible screen. The vertical linearity adjustment serves to adjust the picture for minimum distortion, as when the lower half of a circular object looks flattened while the upper half is egg-shaped. The vertical linearity control is adjusted to make a test pattern circle perfectly round.

Horizontal AFC

While the sync circuits are responsible for the precise timing of the picture reproduction process, timing of horizontal synchronization is more severe and more rigorous than the vertical. It is for this purpose that an additional safeguard is provided in the form of horizontal AFC, or automatic frequency control (block I, Fig. 2-1). Briefly, this is a self-correcting system using both the incoming sync pulse and a feedback system in the receiver to maintain horizontal synchronization at all times.

Horizontal Deflection System

The primary purpose of the horizontal deflection system (block K, Fig. 2-1) is for the left-right movement of the beam. For this reason I shall first describe this function only, although, as we shall see later, this deflection system has a most important secondary function.

Because of the added functions of this portion of the set, as well as because of more critical accuracy demands, the horizontal deflection system contains more tubes than the relatively simple vertical circuit. The first tube, V15, acts as the horizontal AFC (automatic frequency control). It receives the horizontal sync pulse from the sync separator and produces an automatic, precise timing pulse for the horizontal oscillator. Horizontal oscillator and discharge tube V16 performs a dual function: the first is similar to that of V13 in block J; the other being required because of the more rigorous demands of this portion of the TV set.

Following the oscillator is the horizontal amplifier, V17, a heavy-duty, high-power tube which performs, in addition to the obvious function of a horizontal amplifier, that most important second function mentioned perviously. The output of the horizontal amplifier goes to the picture tube for the basic function of deflecting the electron beam from left to right. As such its task is quite comparable to the vertical deflection amplifier just described; however, because of the very important added functions this unit performs, the whole horizontal deflection portion of the TV set is much more complicated and elaborate.

At least two of the tubes—horizontal damper V18 and high-voltage rectifier V19—perform functions entirely different from deflection; however, since these two functions are the result of the basic sweep circuit operation and because of the interdependence of these diverse functions, they are grouped together. Special precautions here include a local fuse (does not affect the rest of the TV set) in some makes for the horizontal circuits only, and a protective enclosure or cage because the very high voltage generated here is a potential safety hazard. Even with the cage, and even with the TV on-off switch in the off position, a severe electric shock and consequent secondary injury may result from carelessness in this area. I shall detail these later.

High-Voltage Cage

The high-voltage (HV) (block L, Fig. 2-1) of any TV set is physically distinct and separated from the open part of the chassis primarily for reasons of safety. It is here that the extremely high potential (25,000 volts in many color TV sets) for the picture tube is developed. While it is true that the energy behind this extra high voltage is too feeble to be lethal, the danger from this source is nonetheless very grave. Even secondary effects, such as the reaction to the shock, a resultant fall, etc., can be extremely serious. Caution labels in the vicinity of the high-voltage cage have their purpose, but a clear understanding of what is involved is even more valuable.

In some cases the cage contains only three tubes, the fourth (oscillator-discharge tube) is left outside the cage for practical considerations. Using the four-tube configuration shown in Fig. 2-5, the functions are as follows: V16B is a sort of intermediate tube between the precisely timed horizontal oscillator tube and the horizontal amplifier, V17. Nontechnically speaking, the horizontal discharge tube serves as a transfer stage for the horizontal sweep voltage from oscillator to amplifier. Horizontal amplifier V17 performs primarily the function of amplifier, which brings the level of the horizontal sweep up to that required by the picture tube. Tubes V18, the damper, and V19, the high-voltage rectifier, are part of what was earlier called the important second function of the horizontal amplifier.

At this point I feel the need to elaborate somewhat on the mechanism of picture painting, as I've called it, in order to make some troubleshooting procedures and corrective adjustments more understandable. The picture tube of any set, monochrome or color, consists of three functionally distinct components: the electron gun, the positioning or deflection structure and the screen.

The gun is physically is located in the rear or neck portion of the tube, in which the glowing heater can be seen when the set is on. It performs the function of generating, shaping and focusing the electron beam to a pencil-point sharpness at the point where it hits the screen. The image brightness adjustment is also connected to this portion of the tube. It may be worthwhile to repeat here that this is an invisible beam of electrons, not light. The light you see on the screen is the result of the electron beam striking a phosphor coating inside the picture tube.

Positioning of the beam, including its zig-zag movement across and down the screen, is accomplished by the outputs of the horizontal and vertical amplifiers connected, respectively, to the horizontal and vertical deflection yoke (coils placed around the picture tube neck). It is, therefore, obvious that if the beam stays on a single horizontal line instead of moving gradually down to create a complete picture, the vertical deflection system is at fault. By logical deduction one might conclude that a failure in the horizontal deflection system would produce a single up-and-down line on the screen. Logical as this may seem, this is not the case, because of that second important function of the horizontal amplifier referred to earlier. We shall now see why this is so.

The third portion of the picture tube, the screen, is an electron-to-light converter. The coating on the inside of the tube has this capability, but it will produce light only if the electron beam strikes the coating at sufficiently high speed. This high speed is imparted to the beam by a high voltage (actually many thousands of volts) generated by the horizontal amplifier under normal operation; therefore, whenever the horizontal amplifier is not performing, no such high voltage is generated and no light whatsoever appears on the screen.

We can now return to the description of the tubes in block L and their functions. As stated previously the horizontal amplifier generates, almost as a by-product, a high voltage ranging up to about 25,000 volts. Since this happens to be an AC voltage—DC voltage is required for the acceleration of the electron beam—a rectifier tube (V19, Fig. 2-5) is used to convert the AC to DC. This may be a 1×2, 1B3, 1A×4, etc. Unlike most glass tubes in a TV set, this one glows very dimly, and cannot easily be seen; however, failure of this tube

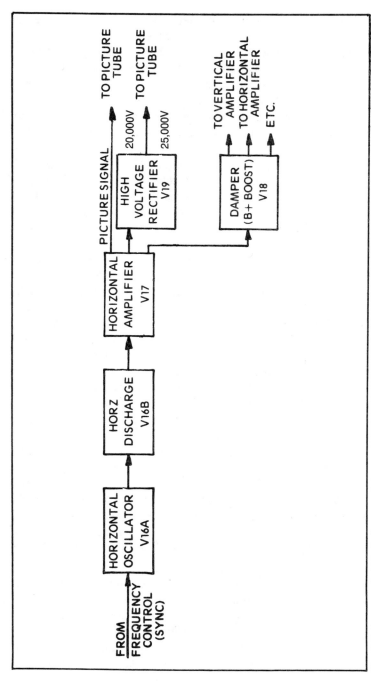

Fig. 2-5. An expanded block diagram of the stages represented by blocks K and L in Fig. 2-1.

alone, even if the horizontal amplifier operates normally, will result in no light whatsoever on the picture tube screen.

The damper tube, V18 serves another useful by-product function. In fact, this tube is sometimes marked on the cabinet chart as *B + Boost*. (B+ is a supply voltage used by all stages in the receiver). Since all tubes in the TV set require a voltage (up to 350 volts or so) for operation, advantage is taken of an available surplus voltage in the horizontal amplifier to boost the nominal 150- and 200-volt supply to about 350 volts where this is needed. The damper or boost tube is also a rectifier, changing AC to the required DC. What is important to the beginner is the fact that if this tube fails, some vital voltages are interrupted and the picture again disappears completely. In fact, the horizontal amplifier which generates this boost voltage is dependent on it for its operation, a sort of pulling-yourself-up-by-your-own-bootstrap scheme that works very well.

Another key device in block L in some sets is a fuse, not for the TV set as a whole, but only for the horizontal deflection system. Failure of this fuse, sometimes without apparent cause, will disable the horizontal amplifier and indirectly the B+ boost and the high-voltage power supply. The symptoms will then be the same as a failure of any other vital link in block L, that is, the loss of all light on the TV screen. Failure of this fuse is rather infrequent, and because it is a part of the rather dangerous high-voltage system, it is not accessible from the outside rear of the cabinet, as is the main 115-volt AC fuse. In addition it is often soldered in place. Fortunately, satisfactory replacement can be made in almost all TV makes without the use of a soldering tool.

There are a few operating adjustments associated with the horizontal deflection system. These adjustments are reasonably accessible, quite safe, and with use of the manufacturer's instructions where available can be easily made. The adjustments are made at the factory and may require minor touching up after long periods (2 years or more) of operation; or they may need readjustments after replacement of a defective part. The adjustments and their functions follow.

Horizontal Width—This is a slotted-shaft adjustment for obtaining the correct picture width on the screen. It is a multi-turn adjustment, and should not be necessary except on rare occasions, as when replacing the horizontal amplifier tube (V17) due to old age. It should not be necessary to turn this adjustment more than four or five full turns either way. If this does not produce the desired result some other defect is probably the cause. Incidentally, turning this adjustment too far in either direction is likely to completely disengage the moving core (slug)—it will actually fall out. If this happens it may be beyond the beginner's ability to recover and replace.

Horizontal Hold—This adjustment is physically very much like the width adjustment. Its purpose is to preset the circuit controlling the movement of the beam across the screen (left to right) to its approximately correct position, so that the automatic frequency control (AFC) can take over and make the timing precisely correct. We shall see later that in some cases of picture tearing a slight adjustment of this circuit will restore the picture to stability. But the horizontal hold control cannot correct any sweep faults. Further, when an adjustment is made it must be done very gradually, sometimes only a fraction of a full turn, for results.

Horizontal Drive—In some older sets an adjustment marker *H-drive*, for horizontal drive, may exist. Hopefully most of these sets have gone to that big trash dump in the sky. This adjustment has the effect of stretching one side of the picture and should only be made when poor horizontal sync exists. Don't try this adjustment unless you really feel that it will help. Most TV servicers only attempt this with proper test equipment; however, there is an alternate method of adjustment that required only a screwdriver and good eyesight.

Tune in a very weak station, one with lots of snow. Turning the adjustment in one direction will produce a white vertical (up and down) line in the center on the screen. Now turn the adjustment until the line just disappears, and you've got it.

Horizontal Lock-In—This adjustment, like the previous one, is found only on some sets. Its behavior is somewhat similar to the hold control described earlier. Turning it from its normal position toward either extreme will produce a diagonal tearing of the picture, in one direction or the other, depending on the direction of rotation of the adjustment. Figure 2-6 is an example of such tearing. The lock-in control's correct setting is the point where channel changing or adjustments will not produce any tearing due to the automatic control by the horizontal sync pulse.

Centering Controls—These functions used to be rear-of-chassis adjustments on TV sets a number of years ago. They are no longer in use today, as they are replaced by a mechanical centering device which is part of the picture tube deflection yoke; however, in those rare cases where one of these adjustments is found on the back of the chassis, the procedure is simple: With a test pattern on the screen (or an equivalent circular emblem of some sort), the control is adjusted for best centering of the image on the screen.

Power Supply

The power supply (block M, Fig. 2-1) is rather simple yet a very important part of the TV set. It supplies power to all the tubes or transistors in the set. A failure here is certain to disable the complete TV set, except, in case of the tubes, it may not prevent them from lighting. The tube, V20, is called a rectifier, and its function is to convert AC to DC, as required by the various tubes and transistors. Sometimes two tubes, side by side, serve the function, in which case failure of one may not completely disable the set; however, since the remaining tube is probably overloaded, it is most prudent to check both and replace both even if one seems to function. Some basic precautions are to be observed when handling rectifiers and tubes, and these *do's* and *don'ts* are detailed in the section covering troubleshooting.

In connection with the power supply it is appropriate to explain the two main types of tube hookups currently used by manufacturers. These are the parallel system and the series system.

Parallel Tube Wiring—With this arrangement all tubes are wired independently of each other, similar to the wiring of a string of modern Christmas treelights in which burnout of one or more does not affect the others. You can tell at a glance which bulbs are burned out. In a TV set using this wiring method it is simpler to locate a defective tube. Should one burn out, it alone will not light (and feel cold to the touch) and is easily identifiable.

An easy way to determine whether the set has parallel tube wiring is to examine the tube chart inside the cabinet. In a parallel tube set all the tube numbers (with two exceptions given below) will *begin* with the same number, usually a *6*. For example, the tubes may be numbered 6BC6, 6CG8, 6BM8, etc. The reason why *6* is a common first number is that they are all 6-volt tubes; in other words all tube filaments operate on 6 volts.

The two exceptions are the picture tube where the first number indicates the tube diameter (21GP4B indicates a 21-inch tube, a 17LP4B stands for a 17-inch tube, etc.), and the high-voltage rectifier, which in the vast majority of black-and-white receivers is a 1-volt tube.

Fig. 2-6. A display on your TV screen similar to this drawing is a case of horizontal tearing.

Series Tube Wiring—This type of tube filament wiring is a characteristic of the so-called transformerless TV sets. As the same implies this TV receiver has no power transformer. There are three ways to identify a series filament string. First, and this never fails, if one tube burns out, all tubes go out, like the *old-fashioned* Christmas tree light string. Even when the set operates normally, pulling out one tube disables all the others. A second identification, also infallible, is the tube numbering system where the first number or numbers of the tubes vary (4BZ8, 10JT8, 38HE7, for example). The third clue is an inscription or imprint on the rear cover of the TV cabinet (or just a printed label) warning that one side of the 115-volt electric line is connected directly to the metal chassis. The requirement stems from a shock hazard that may exist with such construction, a condition exclusive with transformerless receivers.

For troubleshooting a TV receiver with a tube failure a very simple and fairly foolproof test is substitution—replacement of a suspected tube by its exact duplicate known to be good. This can be done most easily in the transformer type or parallel filament set, as will be detailed later. In the transformerless set there is first the problem of locating the suspected tube since all are out when one goes. Substitution is still possible but it is a hit-or-miss procedure in which the defective one may be the last one you try. In a parallel filament (transformer) set you actually can see and feel the tube which burned out; therefore, a different procedure has to be followed in a series string TV set.

Hybrid TV set—As more and more semiconductors (transistor and diodes) are finding their way into TV sets a consideration of a

45

third arrangement—part tubes, part transistors—is of importance to the extent that such sets lend themselves to the efforts of a beginner. There are no 100% solid-state TV sets. At least the picture tube and perhaps one or more of the others are still of the conventional vacuum-tube type. Unlike tubes most of the transistors in today's TV sets cannot be suspected or identified as defective by either light or heat. Transistors have no heaters and operate cold to the touch, except for some power types, which get hot enough to require location on a metallic heat sink to prevent damage due to overheating. Usually, low-power transistors are assembled on subassemblies or boards (sometimes called PC boards) carrying a number of small parts. We examine some examples and outline procedures for dealing with them in section on troubleshooting.

SOLID-STATE TV RECEIVER

Although the block diagram in Fig. 2-1 is representative of practically any TV set it is desirable to familiarize oneself with its solid-state equivalent. This is not only because this type is the standard for sets currently being manufactured, but also because of some significant differences due largely to the nature of semiconductors. A brief explanation of the above statement is in order.

An examination of any block diagram of a tube-type receiver—or the tube layout invariably found inside the TV cabinet—will disclose that many tubes are dual- or even triple-section devices, each section being a stage in itself, all sharing a common glass envelope. Such tubes are easily recognized by their designations, namely, V1A, V1B, V2A, V2B, V2C, etc. In each case the number designates the whole physical assembly, while the letters A, B, and C identify the separate functional tubes in the same glass or metal assembly. Quite often the individual tubes are different from each other and perform different functions, so that although located in one place on the chassis, they connect to different parts of the circuit. Figures 9-1 through 9-4 illustrate this repeatedly. With the advent of semiconductors, however, the picture has changed, primarily because of the nature of transistors; they do not lend themselves to multifunction construction with ease. Thus, while in tube tuners one of the tubes (the mixer-oscillator) is almost invariably a dual-function assembly, the semiconductor counterpart has separate transistors for the mixer and the oscillator. Furthermore, where in tube tuners two tubes suffice, three or more transistors are usually found performing that function in the solid-state receiver. In addition the solid-state receiver often has added stages, consisting of transistors and associated parts, not because of added functions, but rather because of peculiar need for performing the same functions with semiconductors. For these reasons Fig. 2-7 is presented for comparison to Fig. 2-1. We shall briefly compare the two block diagrams for the express purpose of illustrating the nature of semiconductors and their performance from the TV owner's point of view.

There are many multifunction assemblies in the solid-state field, as for example, IC (integrated circuits); in fact, these are exclusive with semiconductors. But these are complete assemblies, such as audio amplifiers, multiple IF amplifiers, etc. They are always part of the same circuit, electrically as well as physically. Such was not the case with vacuum tube construction.

We shall also give an illustration of the application of ICs in suitable portions of the set. This will provide an example of the transition from individual transistor stages to preassembled multitransistor/multidiode entities, which are always replaced as a unit, not only by the most knowledgeable do-it-yourselfer, but even by the most professional repair establishment. These entities are factory assembled and encapsulated.

TUBE-TYPE & SOLID-STATE RECEIVER COMPARISON

In the following comparison the particular function descriptions will not be repeated; they are the same for both block diagrams. We shall however indicate functional differences.

Blocks A1 are the same, transistors Q1, Q2 and Q3 corresponding to tubes V1, V2 and V3; however, while in the tube version V2 and V3 are *generally* a single dual-purpose tube, transistors Q2 and Q3 provide separate functions of mixer and oscillator, respectively.

Blocks A2, the UHF tuner, both use a diode mixer, but the solid-state version adds a transistor UHF oscillator, Q4. Block B, the IF amplifier, in Fig. 2-7 has three transistors, Q5, Q6 and Q7, corresponding to tubes V3, V4 and V5; however, although in many inexpensive tube-type sets there may be only two tubes instead of the three shown in Fig. 2-1, the transistor version is more likely to have three stages, because of the relatively lower cost of the transistors.

The video detectors in blocks C are identical, as diodes have been generally used for this function in tube sets. The video amplifiers in blocks D fully correspond. In both cases the output, which is the video information, goes to the picture tube. Sync information in the transistor version comes from the video detector and not the output of the video amplifier.

There is a slightly different arrangement in case of the sound portion of the signal. In case of the tube set the sound component is taken from the video amplifier, block D, and goes to the sound IF, block E. In the transistor version the sound is taken from the final video IF, amplifier and is fed to a sound IF detector, D3, block E. This strips, or picks off, the sound IF component from the total (composite) signal. It is not a detector in the usual sense, since its output is not yet a sound signal suitable for hearing. The sound IF, block E in the tube version, corresponds to block F in the transistor version, V7 and V8 corresponding to transistors Q9 and Q10, respectively. In some transistor receivers there may be three stages of sound IF amplification, where the particular design requires it for adequate gain. Block F in Fig. 2-1 corresponds exactly with block G in Fig. 2-7; both use a pair of diodes as a discriminator, the real sound detector. Block G of Fig. 2-1 and block H of Fig. 2-7 also correspond, each showing two audio amplifier stages, although in some solid-state receivers there may be as many as four transistors performing the same function as the two tubes of block G, Fig. 2-1. In such a case, the first two transistors might be designated as drivers or preamplifiers while the last two might be a *push-pull* output amplifier.

Block J in the transistor set corresponds to block H in the tube set; all sync functions are accomplished by the two transistors Q13 and 14 corresponding to tubes V11 and V12 in Fig. 2-1. Blocks K, L and M in Fig. 2-7 are the vertical oscillator, vertical driver and vertical output amplifier, corresponding in function to V13 and V14 in the tube version.

In the horizontal deflection circuitry transistor Q18 in block N of Fig. 2-7 performs the same function as V15 in block I, horizontal AFC, in Fig. 2-1. In some transistor TV sets there are usually two diodes associated with Q18. Tube V16 in block U of Fig. 2-1 corresponds in the transistor set to block 0 (horizontal oscillator Q19) and block P (horizontal driver Q20). Block L of Fig. 2-1 corresponds to block Q of Fig. 2-7. Horizontal amplifier V17 is replaced by Q21, damper V18 by D6 and high-voltage rectifier V19 by semiconductor diode D7. In both cases these three are shown in one group since they actually work that way. Block M of Fig. 2-1, shown to contain rectifier V20, or two diodes, corresponds to block R in Fig. 2-7, containing diode D8. Different designs may call here for two and

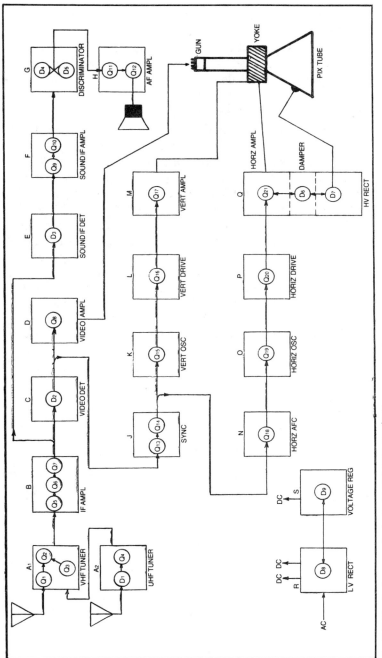

Fig. 2-7. Block diagram of a typical solid-state receiver.

even four diodes, all for the same purpose—rectification of the AC. Finally, block S in Fig. 2-7 has no counterpart in the tube type set of Fig. 2-1, although many tube sets have the same function. This is a low-voltage regulator stage, usually consisting of a diode (D9) or a transistor followed by a diode.

INTEGRATED CIRCUIT APPLICATIONS

In Chapter 5 I recommend that the TV owner purchase the complete package of data for his or her particular model receiver. In such a package there always are schematic diagrams and other partial diagrams necessary for servicing the receiver. Since many modern receivers contain ICs it behoves me to acquaint you with these devices and the manner of their representation on the various diagrams so that they may be understood. Briefly, an integrated circuit is an assembly of various parts, including transistor and diodes, so that only the necessary connections are externally accessible; all other interconnections are internal and permanent. Although this may at first seem uneconomical, requiring replacement of a complex assembly just because one item has failed, actually this is not the case. Preassembly and encapsulation saves more in the long run, making it more economical (and often better performing) than multiplicity of discrete parts individually interconnected.

Chapter 3
Color TV & How It Works

Up to this point color TV has not been included in our discussion, but everything that has been pertains equally to color. In fact, when I discuss the circuits and functions peculiar to color receivers only, you will see that these circuits and functions are additional to the basic TV set circuits and necessary to adjust one control (color) to its minimum (equivalent to off) position in any color TV set to arrive at a black-and-white representation insofar as results are concerned.

Generally speaking all the TV circuits up to and including the video detector-amplifier are common to all signals—sound, video and color.

NATURE OF THE VISUAL SPECTRUM

Figure 3-1 is a graphic depiction of the color components of visible (white) light after passing through an optical device called a prism. Although discrete color blocks are indicated, all colors gradually blend into each other. Figure 3-2 is a much simplified version of the same phenomenon, more familiar to the student of elementary painting. The three primary colors produce intermediate colors as well as white. In the present color TV system, this basic three-color system is utilized to produce the color TV picture as we know it.

COLOR TRANSMITTER

Figure 3-3 is a functional or block diagram illustrating in much simplified form the components of a color picture transmission sys-

Fig. 3-1. Pure white light is composed of all the colors depicted here.

tem. A comparison with the monochrome TV transmitter in Fig. 1-4 shows the basic difference between the two. Let's consider the components of a color system.

Camera Tubes

Three camera tubes (block A Fig. 3-3), one for each of the primary colors together with their individual amplifiers, combine to produce the three components of color information required to produce a color picture on the home TV set. These three components are:

Picture Brightness Information—This establishes the overall color picture brightness in the TV set, not just the brightness of any particular color but the background brightness level (proportionately) correct for all colors.

Picture Color Information—This is the heart of the color signal and contains all the information required by a TV set to reproduce the entire range of colors, corresponding to the original scene.

Color Sync Information—This corresponds (and is in addition) to the earlier described sync functions in black-and-white television. It is sometimes referred to as the 3.58 MHz color burst.

Modulators

The modulator or combiner portion of the color TV transmitter corresponds to the similar functions in the monochrome transmitter. Here the various electrical components of the color picture are combined (block B, Fig. 3-3), without losing their identities, into an overall color signal, which in turn is combined with the sound signal.

Final Amplifier & Antenna

The final amplifier and antenna system are essentially the same as the black-and-white transmitter in Fig. 1-4, except for the addition of the color information. Neither the final amplifier nor the transmitting antenna "know" the difference; that is, they amplify and radiate (respectively) a total television signal made up of the required components. This same is true with a receiver. We shall now proceed to examine a color TV receiver on a function-by-function comparison with its Black & White counterparts.

COLOR RECEIVER

The color TV receiver is a combination of a basic monochrome receiver plus an add-on section which concerns itself only with the color functions of the set. In fact, it will soon become apparent that this add-on section picks up where the black-and-white portion of the set leaves off. Consequently, and as was briefly stated earlier, the standard procedures for analyzing and troubleshooting a color TV set is to first consider it as a black-and-white set. This is easily done by reducing the color signal (turning the chroma control) to zero, thus leaving a black-and-white picture as would be received by a monochrome receiver tuned in to a color program. It is for this

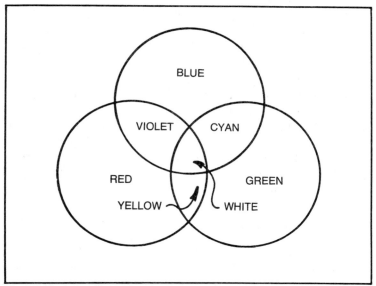

Fig. 3-2. The color wheel principle shows how the three primary colors blend into three intermediate colors and white.

reason also that the block diagram to follow (Fig. 3-4) is divided into two sections by a dashed line. The black-and-white portion is readily recognizable as the familiar monochrome TV set shown in Fig. 2-1; therefore, the black-and-white section receives only a minimum of emphasis in this discussion.

It will be advantageous, however, to view the complete TV receiver as a unit, in order to clearly visualize the continuity between and the transition from the combined monochrome and color functions to the strictly color portions and functions. The overall functional diagram is represented with this purpose in mind.

Figure 3-4 gives a simple, nontechnical means of comparison between a monochrome TV set and the corresponding color set. It shows a complete color TV set, with no reference to black-and-white operation; however, it is obvious that the complete black-and-white TV set of Fig. 2-1 appears with hardly any change inside the framework of the color set diagrams.

In addition to the basic facts of structure this graphic comparison also shows, in greatly simplified form of course, the two main constituents of the color picture. Function blocks O and P are responsible for the three component colors plus the overall color picture brightness information, while function blocks M and N provide the information for the proper color registration and mixing of the three primary colors into the complete range of color shading which makes up the final color image on the home TV screen.

Before discussing individual tube, stage or functional blocks of the complete color TV set let us briefly analyze the complete sequence of functions as indicated in Fig. 3-4 in order to get a concise overall story of what happens. The TV signal, containing picture information and sound information, is intercepted by the antenna. It is a composite (all-in-one) signal. Block A (the tuner) selects any one desired station from the number of stations available in a particular location. The IF (intermediate frequency) amplifiers (block B) amplify the selected station to the signal level required for further processing.

So far the composite signal picked up by the antenna and selected by the tuner has been made larger as required, but it still is riding piggyback on a carrier. Block C, the video detector, performs but one simple main function—it separates the intelligence (picture, sound and sync information (from the carrier, which served its purpose as a vehicle for transmission of the signals). The output of the video detector is the composite signal described previously.

Although the carrier appears to be discarded at this point it is not. As far as the intelligence goes it has done its job. Nevertheless it

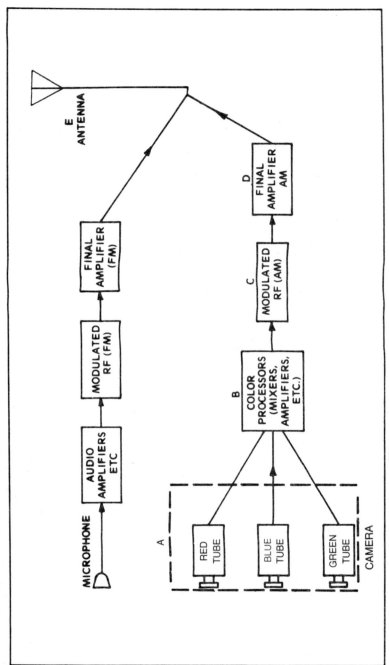

Fig. 3-3. Simplified block diagram of a color TV transmission system.

is utilized to perform a vital function of adjusting the TV receiver's amplification to the required level; that is, the weaker the incoming signal, the more it is amplified, and vice versa. This is known as automatic gain control (AGC for short). Incidentally, the AGC adjustment is a back-of-the-set adjustment that sometimes required resetting on both color and black-and-white receivers.

Block D in Fig. 3-4 is the video amplifier. True to its name, it amplifies the video signal together with its other components (sound and sync signals) received from the video detector just described. At the output of the video amplifier the three signal components separate, each going its own way.

Picture information (commonly called video) goes to the picture tube for conversion into an optical image. Sound information, a sound-modulated carrier (sound energy piggybacked on an RF carrier), goes to block E—the sound IF amplifier. The signal now is the equivalent of an FM radio signal in an ordinary FM receiver. As such it is a sound signal superimposed on a carrier. This in no way contradicts our earlier statement that a carrier has been discarded by the video detector, although the technical explanation is of no consequence to our purpose. After amplification the signal goes to an FM detector (block F) or discriminator where sound energy only is extracted. Block G is an audio amplifier, which brings the sound energy to the level required by a speaker connected to the output of block G.

At this point you may be wondering why the same amplifiers appear over and over again, interrupted by other functions in the chain. Wouldn't it be better to have just one amplifier in one place and do all the required amplification at once? It would seem simpler; however, a TV (and radio, too) receiver is designed to receive a very faint signal from the antenna. In fact, such a signal may be a million or more times weaker than the voltage of the ordinary flashlight cell (nominally 1.5 volts). Necessary incidental losses within the TV set make this situation even more critical. In other words the incoming TV signal must be amplified millions of times before it becomes the final product (picture or sound).

The most practical and most (technically) efficient way to achieve this tremendous magnification is to do it in small stages and at particular locations in the set. Thus, some initial amplification (RF amplifier) is done immediately after the signal arrives from the antenna. Additional amplification is provided after initial processing (conversion to IF in the tuner), this time in the IF amplifiers. A third and usually final portion of amplification is accomplished, after another conversion (detection), in the video amplifier for the picture signal and in the audio amplifiers for the sound signal. There also are other amplifiers in the TV set, for such nonsignal functions as sync and AGC (automatic gain control).

Meanwhile, back at the video amplifier the third component of the composite signal from the video amplifier is the synchronizing-timing signal (called sync for short). Block H contains sync amplifiers, separators and shapers, with an overall function of making the timing pulses suitable for controlling the accuracy of the sweep voltages via the horizontal and vertical oscillators. Block J is the combined vertical oscillator/amplifier which produces the voltage for vertical deflection of the picture tube beam. Block K is the corresponding group for horizontal deflection, as detailed earlier for monochrome. The dotted line in block K separates the strictly sweep function from the incidental (although most important) functions of B+ boost and high-voltage generation. Finally, block L represents the power supply for the entire TV set, including some adjustments and controls for the picture tube.

So far I have described a complete black-and-white receiver, although hidden within its various stages were color signals—information convertible to a color picture after proper processing. The following paragraphs describe such conversion and processing.

From the video amplifier (block D, Fig. 3-4) the two components of the color signal go to the color-only portions of the receiver. Color picture information goes to block O, a color amplifier. After the signal is amplified to the required level, it goes to block P, the color detector or demodulator, then to the individual color amplifiers in block Q. Here there are three separate amplifiers, one for each of the primary colors, whose outputs are applied, respectively, to each element the three-color gun in the picture tube.

The second component of the color signal is the *burst* information. It is responsible for the proper color registration in the picture. It is fed to block M, the burst amplifier. From here it is used to synchronize the color oscillator for precise timing or synchronizing of this all-important component of the color process. While block M of the color section is analogous to block H (sync circuits) in the black-and-white portion of the set, the color oscillator is peculiar to the color process. No corresponding function exists in the black-and-white receiver.

The color oscillator signal (3.58 MHz) is used in the color demodulator (block P) to retrieve the three primary color signals. If

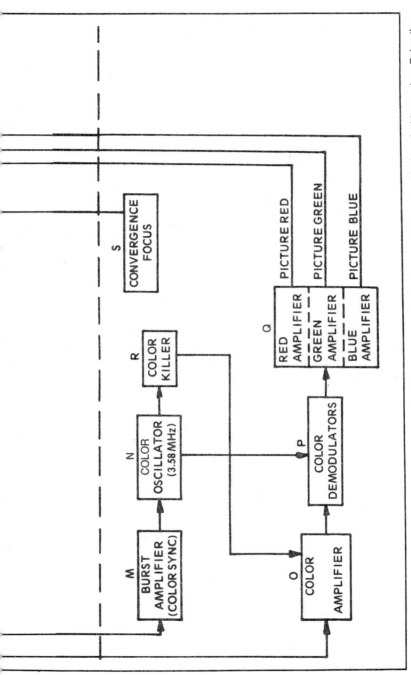

Fig. 3-4. Functional block diagram of a color receiver. Those functions above the dashed line parallel those in a black-and-white receiver. Below the dashed line are color-only functions.

each color is to be reproduced accurately the color oscillator must operate in step with a similar oscillator at the transmitter. A failure in the color oscillator causes a complete loss of the color in a picture. The color oscillator output is used as a carrier (just like a picture or sound carrier) to detect or demodulate the color signals. The color signal is called a subcarrier because it is within the black-and-white picture carrier.

There are three additional functions in the strictly color portion of Fig. 3-4, namely, the color killer and the focusing and convergence systems (blocks R and S). Of necessity, I shall leave the explanation of these for later.

COLOR PICTURE TUBE

To better understand the construction and operation of a color picture tube, let's quickly review the features of a monochrome picture tube.

A monochrome TV picture tube consists of three main sections: the gun, the deflection system, and the screen. The gun produces a stream of electrons, which, with the aid of properly applied voltages, is focused to a sharp point on the face of the tube (the screen). The deflection system, again by virtue of correctly applied voltages of proper characteristics (horizontal and vertical sweep voltages), moves the beam to the desired points, in proper sequence, on the screen.

Now, if instead of a black-and-white phosphor (coating on the inside surface of the TV screen), we substituted a red phosphor (or a blue, or a green), the same exact picture tube just described would produce the same TV picture, but in red (or blue, or green), with all the various gradations in brightness as in the black-and-white picture. It is a bit of oversimplification, but nonetheless true, that the color TV picture tube is a combination of three tubes, a red tube, a blue tube and a green tube in one glass envelope; however, while this picture tube has three distinct guns, each emitting an electron beam for each color, it has but one screen, for all three colors. This accounts for the great complexity of color TV tubes. In order to understand some vital adjustments and corrections that are sometimes necessary on a color TV set it is first necessary to understand the structure and functioning of a color picture tube.

In contrast with the black-and-white picture tube screen, which has a continuous coating of phosphor and where the electron beam may be placed on any point on the screen, the color tube screen does not have a continuous coating but consists of a large number (hundreds of thousands) of color dots, and the electron beam must hit the dots only. Specifically, the screen is composed of three-dot groups; each group has a red, a blue and a green dot closely spaced but not quite touching. A recent improvement in color picture tubes has been realized by putting a black border, a nonluminous coating around each color dot, to reduce unintentional blending of the colors. The dot arrangement is alternated in each group so that no two adjacent dots in any direction are of the same color. For better visualization, consider the dot groups as subminiature billiard balls, arranged in triangles (three in a group, almost touching) with no two balls in the triangle being the same. Figure 3-5 shows the dot layout on a portion of a color screen as just described.

Immediately behind the tricolor picture tube faceplate is a perforated metallic plate very precisely arranged and positioned so that each hole in the perforated plate lies in an exact position behind a group of color dots, a color triad, on the faceplate. When the plate is correctly positioned, the red beam goes through the hole and strikes the red dot only. Similarly, the blue and green beams passing through the same hole will strike the blue and green dots, respectively. This is a precision structural alignment and is independent of the characteristics of behavior of the TV set.

For the sake of clarity let me state again that the three segments of the electron gun in the tricolor picture tube are identical. They are called red, blue, and green only because of their physical positioning with regard to the aperture mask (the plate behind the screen) and faceplate not because of any color difference between

Fig. 3-5. This sketch illustrates the dot pattern found on a color TV picture tube screen.

them. All three produce identical electron beams with no inherent color characteristics.

It should be apparent from the above description that for the proper excitation of the three colors the three electron beams originating at different positions in the neck of the tube (see Fig. 3-6) must be made to properly bend or *converge* so as to go through the holes in the aperture mask instead of striking the space between the holes. While this is approximately provided by virtue of tube design, it is not sufficiently accurate for a satisfactory color picture reproduction. To insure this special convergence devices are employed. There are two such sets of convergence adjustments. One consists of a set of magnets placed around the neck of the picture tube and adjustable for physical position. This set is called *static convergence* controls and is best left alone. The second set of convergence adjustments is of the rear-of-the-set shaft types. Three groups of this type of adjustment controls exist, one for each of the three primary colors.

While focusing is a function common to all TV picture tubes it is much more important in a color tube due to the strict compatibility requirements between the three electron beams, the hole in the aperture mask and the color triad just in front of each hole.

Block S of Fig. 3-4 indicates the convergence function just described. Although the convergence adjustment shafts are accessible at the rear of the receiver, these adjustments require extreme care as well as complex equipment in order to set them properly. Such is not the case with the focusing adjustments. With proper care (as I shall indicate under troubleshooting) this adjustment can be satisfactorily performed by a beginner.

Incidentally, at least one color TV manufacturer has built in a convergence tests accessory, making it relatively simple to carry out these adjustments when required. The trend in the industry seems to be toward inclusion of the simpler self-test or self-correct capabilities in the receiver. A common example is the demagnetization or degaussing coil which prevents and corrects color deterioration due to stray magnetic influences in the immediate vicinity of the color set. Prior to this the performance of a color TV set, even though it was in normal operating condition, was affected (sometimes seriously degraded) by the appliances or electric wiring in the immediate vicinity.

A third strictly color function mentioned earlier is the *color killer*, and it is described fully later. At this point suffice it to say that the purpose of this is to make sure that no color appears on the screen during a black-and-white picture.

RECENT COLOR TUBE IMPROVEMENTS

In Fig. 3-6 I illustrated the basic color tube using the three component colors in a triad arrangement, relying on the properly perforated plate or *shadow mask* behind the screen to converge the three component beams as required to produce the desired color. Even with a so-called border guard—a black, opaque area separating the color dots—and with improvements in convergence techniques, this construction left much to be desired from the viewpoint of color purity and picture fidelity. Subsequent to this two significant improvements were made in color tubes so that now many TV sets have one or the other of these improved tubes.

The first of these was made a few years ago by one of the foreign manufacturers and still is used. This tube is known commercially as the *Trinitron* produced by Sony Corporation. This uses a three-beam gun, with a color faceplate consisting not of triple dots, but of vertical strips. These strips or stripes are also separated by opaque strips to prevent unintended color overrun or blending. The

Fig. 3-6. Sketches comparing the gun structure. (A) Black-and-white gun. (B) Color picture tube gun.

strips or stripes are continuous from top to bottom. Figure 3-7 shows in sketch form, the gun, focusing and faceplate structure of this tube. The Trinitron equivalent of the shadow mask, called the aperture grille, which has vertical slits instead of triad holes, is claimed to be more accurately producible and to provide a greater beam converage of the color screen. In a properly adjusted set of this type, that is, when the brightness, contrast and color intensity are normal, it is possible to actually see each of the vertical color lines by getting close to the screen. The appearance resembles that of a fine cloth weave. In addition to the superior picture quality claimed for this type of tube there are also the advantages of simplified convergence adjustments, as well as fewer adjustments compared to the basic original color tube described earlier.

The second type of improved color tube, and one found in more and more use by a number of manufacturers, also has a vertical slot structure, but, unlike the Trinitron continuous stripe arrangement, it consists of slots separated by opaque areas. In addition these types employ a three-gun in-line arrangement in contrast to the triad gun arrangement and the single-gun three-beam type. It is claimed that this type of shadow mask also provides for better color distribution and depth. Both of these types, the Trinitron and the slotted-

Fig. 3-7. Sketches showing construction of the Sony Trinitron picture tube (courtesy Sony Corporation).

aperture type, are superior to the old, three-dot type of aperture structure. In addition, and perhaps only incidentally, these tubes are also shorter because of the wider angle encompassed by the screen compared to earlier types. While this may not be of particular interest to the TV owner, other than perhaps the decreased depth of the cabinet, there are some definite advantages from a technical viewpoint also. While there still are a few sets with the three-dot matrix type of picture tube, they are said to have been improved with regard to focusing and color depth or brightness.

There is no simple way to identify the tube type from casual viewing by the observer. But all the manufacturers describe their products so as to leave no doubt about these features. Thus, Trinitron is a sure identification of the continuous-strip mask type, the in-line term identifies the slotted-mask tube (there may be other additional terms used in advertising), and the improved three-dot types are identified by their manufacturers as the mask-focus type, suggesting that each mask hole acts as a focusing lens for sharper, brighter images.

There is one other feature of some of these newer picture tubes worth mentioning, although technically of minor value. This is a mild light filter (picture tube seems dark when the set is off), of a smoky hue, intended as an antireflection device, so that the face of the tube will not act as a mirror and reflect objects in the room.

Chapter 4
Color-Only Sections

The best place to begin an examination of those sections of a color receiver that deal solely with color reproduction is the video amplifier since that's where differences begin. Figure 4-1 is a detailed block diagram of the color-only sections of a hypothetical TV set. I do not mean an unreal TV set, it is a diagram of a color section most suited for our purpose, that is, an illustration of functions, adjustments and control locations of potential malfunctions. It consists entirely of the functional blocks found in actual color TV sets.

Beginning with block A the video amplifier of the black-and-white portion of the set, we see that the color signal together with its timing (burst or sync) signal go to block E on the color side of the dashed dividing line. The dashed line from block A to block F is an alternate way for the color timing signal to be taken out of the video amplifier; thus, in some sets, two separate signals may go from the video amplifier to the color portion of the set—the picture information to block E and the burst information to block F. In more recent TV sets, the combined signal is amplified in the first color amplifier; then the burst separates and goes to the burst amplifier (block F) while the color (chroma) signal goes to the second color amplifier, block G. Here again some TV sets have only one color amplifier, so that block G would be missing. Whether a particular TV set has one or two *chroma* amplifiers, the front-of-the-set control marked color is located here and it adjusts the color content of the picture from all black-and-white (no color) to full color (and beyond, if turned to its extreme position).

BURST AMPLIFIER

The simplest way of describing the function of a burst amplifier (block F in Fig. 4-1) is to say it is exactly the same as the function of the sync amplifier (block H, Fig. 3-4). Input to the burst amplifier is derived either directly from the video amplifier (block A, dashed arrow) or via additional amplication in the first chroma amplifier (block E, solid arrow). The output of the burst amplifier is ultimately applied to the 3.58MHz oscillator (block J) for precise control (timing) of the oscillator.

COLOR SYNC & OSCILLATOR CONTROL

The color sync phase detector (block H) and oscillator control (block I) are grouped together for two reasons: first, both perform functions incident to controlling the accuracy of the color oscillator, and second, they may be considered, in a sense, optional. While some manufacturers use this function for oscillator control, others achieve the same result in a different way and do not incorporate these two in the TV receiver design. What is important to the TV set owner is the fact that when these functions are found in a set, they are links in a chain, so that if one or both should fail, the 3.58 MHz oscillator may wander sufficiently to cause some serious deficiencies in the color rendition and balances (by cutting off some of the color frequency components in the output signals).

As we shall soon see the 3.58 MHz oscillator is one of the ingredients in the conversion of the chroma signal to the red, blue and green portions of the final color picture. The balance between these three primary colors is, of course, essential for a normal color picture and can be achieved only when the 3.58 MHz oscillator signal is fed to the demodulators exactly in time with the corresponding signal in the transmitter. Control circuits and tubes in blocks H and I establish and maintain synchronization between the transmitter and the receiver, and a failure here invariably degrades if not altogether destroys the normal color picture.

COLOR OSCILLATOR

The 3.58 MHz precision oscillator (block J, Fig. 4-1) is a crystal-controlled oscillator, which is basically very accurate and stable even before being further controlled by the burst signal from

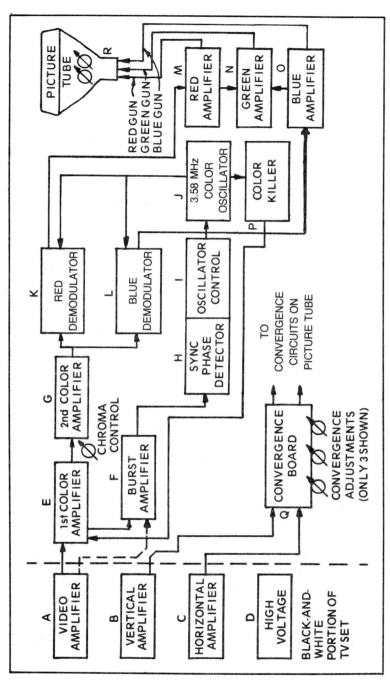

Fig. 4-1. Detailed functional diagram of the color-only portions of a color receiver.

the transmitter. An examination of the immediate vicinity will locate the crystal. It is a fragile, tissue-thin wafer of a synthetic, quartz-like material (the natural mineral quartz is seldom used nowadays) encased in a hermetically sealed plug-in unit, as illustrated in Fig. 4-2. Although the little box requires careful handling, it is relatively durable; however, occasionally it may become intermittent (stop functioning, then start again after the TV set is switched off and on) or it can fail completely. In either case, if the crystal malfunctions the color, but not the picture, disappears.

COLOR DEMODULATORS

Blocks K and L are the color demodulators. As shown in Fig. 4-1 the combined color signal comes from the (second) chroma (color) amplifier, while the second input is from the precision-controlled 3.58 MHz oscillator (or reference signal). Notice that there are only two demodulators, one for the blue signal, the other for the red. The green demodulator, as such, does not exist, and it is unnecessary. Since a combination of the primary colors, blue and red, will produce green, no separate demodulator is required. The green signal is derived by mixing a proportional amount of red and blue.

The function of the color demodulators, stated in nontechnical terms, is to extract the color information from the combined color-plus-carrier signal. The color demodulator is a detector. While two demodulators are shown, it does not necessarily mean that every TV set will have two tubes performing these functions; some may use two transistors for this purpose (in the most recent sets), while others may use a dual tube in a single envelope. Whichever the case, the function is the same.

In case of malfunction a failure of a two-in-one tube will completely shut off all color. Should one section of such a dual tube fail (this is not unusual) or should one of the two demodulators fail in sets with two separate demodulators, two of the three colors will be missing. This is due to the fact, as mentioned earlier, that red and blue color components are combined to produce the green primary color as well as all the hues of green.

COLOR AMPLIFIERS

Blocks M, N and O—the final color amplifiers, one for each of the primary colors, are similar to the video amplifier of a monochrome TV set. Since each color component signal has its own

Fig. 4-2. The crystal used to control the frequency of the 3.58 MHz oscillator is usually hermetically sealed in a metal container. Some are designed to plug in, others are soldered in.

independent path from the demodulator to the picture tube, a failure in a particular color (or blend of that color) is unambiguously traced to the specific amplifier. As I mentioned earlier the block diagram Fig. 4-1 is not of any particular manufacturer's TV set; hence, differences are possible.

COLOR KILLER

Block P represents the color killer, a violent title that means just what it says. The function of this section is to prevent any color from showing on the screen when a black-and-white picture is being received. To begin with let us remind ourselves that a monochrome picture in a color TV set is produced from the correct combination of all three primary colors. In other words, a black-and-white picture on the screen comes not from no color, but from all colors in the exact proportion, just as white is not the absence of color but the presence of all colors.

Secondly, and this follows, that to transmit a black-and-white picture requires proper balance of all three primary color components. While balance would assure no coloration of a black-and-white picture, such balance cannot be taken for granted. Consequently, color fringes may appear on the color TV set when they are not supposed to appear, that is, when the picture is transmitted in monochrome.

It should be remembered that in all cases, whether monochrome or color pictures are transmitted, the picture tube receives a brightness signal which is actually the picture minus the color. Of course, when no color is transmitted this brightness signal is the picture while the chroma amplifiers have nothing to contribute. Similarly, when a color picture is received on a black-and-white set (in which there obviously are no color amplifiers) this brightness signal is all there is and it produces a black-and-white picture without any special adjustments or color killers.

Finally, it must be remembered that the color amplifiers as well as the red, blue and green guns of the picture tube operate on current (electron) beams, not on beams of colored light; thus, any current in the path of a chroma amplifier (which automatically feeds one of the three color guns) is capable of exciting the color dots on the picture tube screen regardless of whether or not color is being transmitted. To prevent this unwanted color the color killer function is required.

In operation, the color killer circuit keeps the chroma amplifiers in a nonamplifying condition (or cutoff) until a color picture is received. At such time, a signal from the color sync (burst signal) disables the color killer, releasing the chroma amplifiers to perform their normal functions.

CONVERGENCE

Convergence (block Q) is mostly a problem with three-gun picture tubes. In the old-fashioned black-and-white picture tube, or for that matter, in the Trinitron color tube in some imported, the problem of convergence is greatly simplified. Convergence means aiming two or more beams so that they converge and meet at one and the same spot. As will be recalled from the description of the three-gun color tube the aperture mask behind the color screen has one pinhole behind each group of three color dots (triads). Not only must the three electron beams go through the same pinhole, but they must be so positioned and directed that the red beam, after passing through the aperture, strikes only the red color dot, and not one of the other two. Similarly, the blue or green gun must be so oriented that only the blue or the green dot is excited. It is the critical function of the convergence circuits and adjustments to insure this if true color separation is to be realized.

It should be noted here that the basic compatibility between the three color guns, the single shadow mask aperture and the three color dots just in front of this aperture is insured by the original construction of color tubes. This, however, does not obviate "live" in-the-receiver adjustments and corrections.

There are two types of convergence adjustments in current three-gun color TV sets. One is called static convergence, the other dynamic convergence.

Static convergence of the three beams is a mechanical adjustment of the positions of small magnets on the neck of the picture tube. Its purpose is to have the three beams converge at the center of the screen only. Stated simply, the purpose of static convergence is to insure that the three guns, starting (necessarily) from three different positions at the back end of the tube, produce one and only one spot when pointed to the center of the screen. This is the correct starting point for dynamic convergence.

The need for dynamic convergence stems from the fact that the three beams travel different (longer) distances when going to the far corners or edges of the tube than when going to the center of the screen. Figure 4-3 shows what a crosshatch pattern would look like as a result of the difference in the paths the beams travel. Furthermore, since the length of the path from the gun to the screen is different for each spot on the screen (the shortest being to the center, the longest to any of the corners) it is necessary to continuously adjust the convergence forces as the beams sweep the screen. This is accomplished by dynamic convergence, that is, continuous convergence correction during sweep.

In most color TV sets all dynamic convergence circuits and adjustments are grouped together, usually on a discrete circuit board, often called the convergence board. While the required continuous correction takes place electronically, it is first necessary to preset a number of controls (two or more for each color) so that the automatic electronic correction will be just right. In other words, dynamic convergence adjustments are intended to preset initial conditions from which the automatic circuitry takes over.

Due to the complexity of convergence adjustments the procedure may not be within the ability of the majority of beginners to perform. This is due not so much to the technical difficulty as it is to the need for some very specialized test equipment. There is at least one TV manufacturer, however, who incorporates such "specialized test equipment" into its TV sets and provides lucid instructions for convergence adjustments. But, even for the majority who will not be able to perform this function themselves, understanding the whys and hows will aid them in understanding what a professional TV servicer is doing.

In illustrating the phenomenon of nonconvergence of Fig. 4-3 I show a crosshatch pattern. In Fig. 4-3A the lines are essentially straight to the very ends, while in Fig. 4-3B noticeable curvature appears and increases the farther the line moves from the center, both in the horizontal and vertical directions. In an actual color picture improper convergence will be accompanied by some color distortion and fuzzing near the ends of the lines. But for proper examination and correction, a test instrument capable of producing a crosshatch pattern on the screen, as in Fig. 4-3, is required. Such is the specialized test equipment referred to earlier.

The convergence procedure involves a sequence of adjustments of convergence board controls while observing a crosshatch pattern (from the test set) on the screen. Each color and each position (left, right, top and bottom) are individually adjusted until the overall crosshatch is as linear as possible. Some sets have two sets of controls, one for blue, the other for red and green combined.

From the viewpoint of the beginner it should be added that convergence problems are neither frequent nor chronic. They may be required when a picture tube is replaced, but seldom otherwise; however, should the need be visually apparent, you know what is needed. I mentioned earlier that one of the color TV sets on the market has such a color signal generator built in. It should be added that the adjustment procedures for this set are quite specific, as they should be, since they serve a particular set and can point to each control by symbol and location, and outline step-by-step procedures for the complete convergence adjustments for this particular set only, of course.

FIXED STATIC CONVERGENCE

For some time now improvements in picture tube design as well as in deflection techniques have made static convergence obsolete so to speak; thus, the starting point for convergence adjustments is the dynamic type—no static adjustments are required. In addition, dynamic convergence has been simplified. Since these changes are very much peculiar to each set, no general procedures can be given; the service information for each particular model should be followed for detailed procedures.

COLOR PICTURE TUBE CIRCUITS

Block R in our typical color chassis (Fig. 4-1) is the color picture tube and its associated controls. There are a number of adjustments and controls, usually two for each color gun, although the red gun may sometimes have only one. There are other adjustments associated with the picture tube, but these are on the yoke and other assemblies on the tube neck, which I return to later.

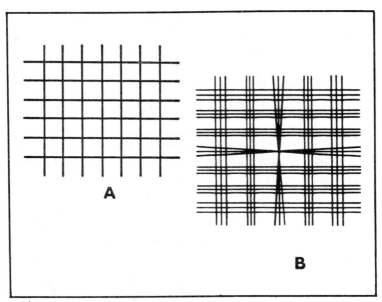

Fig. 4-3. Crosshatch pattern A shows ideal convergence, both at the center (static) and corners (dynamic). Pattern B shows good static convergence but misconvergence at the edges.

At this point we concern ourselves with the controls relating to the color aspects of the TV set only. The two groups of adjustments are the screen and drive adjustments. In most receivers, there is a screen adjustment for each color—red screen, blue screen and green screen; however, there may be only two drive adjustments, one for blue, the second for green. The design of the more recent color tubes and receivers often eliminates the need for the red drive adjustment.

In spite of the fact that these controls, whether six, or five, or even fewer in some sets, are marked with particular color names, their adjustments are best made for a balanced black-and-white picture. These adjustments are often referred to as the *gray-scale* adjustments, meaning that they are set so that the complete range or scale of light is reproduced on a monochrome scene. Once this is done the color picture should require little, if any, touching up for best color balance.

As to the basic need for these adjustments, this stems from the fact that the three-gun picture tube is in fact three fairly independent tubes with some variations in characteristics between them. Adjustments are intended to compensate for and equalize these differences, as well as for some differences in the three signals reaching these three guns. Furthermore, since the correct proportions of the three colors are required to achieve a pure black-and-white picture, correct adjustment for such a picture, therefore, implies that the color balance is correct.

Screen and drive adjustments, although interdependent, can nevertheless be made one set at a time. First, the receiver brightness control is set for a fairly middle-to-dim level with the color control to minimum. Next, each screen control is advanced to the point where a trace of color appears, then backed off to a point just beyond where this trace of color disappears. Next, the brightness control is advanced to what is considered normal brightness and the drive controls adjusted for a full gradation from dark to white on the screen, without any pitch black or glaring white patches showing. In other words the complete range of light to dark should exist. A second touchup of the screen controls may be required, depending on the end results.

If this is carefully carried out there should be good color balance when a color picture is received, although some final touching up may be indicated if the color does not seem to be balanced. To the beginner a balanced color picture is best described as what seems to the eye to be a most natural picture.

Perhaps it might help to indicate what is not a balanced color picture. If, for example, a color picture has a bluish hue (or greenish or reddish), regardless of whether it is sky, or skin, or grass, obviously there is an excess of blue (or green, or red, as the case may be) in the picture. It is actually possible to create such an artificially tinted picture by advancing one of the screen controls and observing the change.

A word of caution is in order, however. If you want to make the test, you can best insure restoration of the color balance to its original state by observing, and perhaps marking, the position of the control about to be changed. Then is it only necessary to reset the control to its original position to obtain the original color balance.

Incidental to the discussion of color balance it is appropriate to comment on artificial settings of the color balance circuits for the sake of particular or special effects. It should be remembered that the color TV set, like a painter, starts with three primary colors, then mixes them in proper proportions to obtain any desired color; however, while the artist in order to achieve the exact hue or tint may have at his or her disposl many more than the three basic colors, not so the color TV tube. Unlike the artist's pallette, there are only three colors to start with. This necessarily limits the versatility of the color tube. While there are many other limiting factors, this is the major one.

For best overall results the proportion of the three colors is such that all white or all blue or all green are easiest to reproduce, while gradations and shadings are not so easy to reproduce accurately. In practice the eye is very tolerant on color shades, except on such tints as flesh tones. While it is possible to adjust the color balance of a set to favor flesh tones, such favoring is usually achieved at the expense of color balance; that is, the general background (or some not so intended objects) will take on the characteristic flesh tones. The clue to the problem lies in the favoring. If for technical (design transmission, etc.) reasons flesh tones do not look natural on a properly adjusted color receiver misadjusting the color balance circuits to favor the flesh tones will invariably distort the color balance on most other tints. It is like misadjusting the tone control on a radio receiver or phonograph to obtain better bass (low-frequency) response by cutting off the high and medium frequencies essential to balanced sound.

Chapter 5
Introduction to Troubleshooting

It is not our intention to tell a beginner in TV repair what to do without also attempting to explain why something is to be done, or why a certain procedure should be followed. Unless you first understand what a tube, a transistor or an assembly is supposed to do, basically, your chance of success in effecting a correction or repair is much smaller than it could be. That is also what I mean by potentially unsuccessful *fixes*.

While it is possible for you to look on a chart of symptoms and find the one that seems most like that in your TV set, then follow the suggested remedy and actually get the desired result, the likelihood of such luck is rather small; however, your luck will improve in direct proportion to your familiarity with your set and your understanding of how it functions when operating normally. It is not impossible for you to look up a fault in the index, which will refer you to a certain page of this book, and by following the instructions end up with a *fixed* set. But your chances of continued success are better if you thoroughly acquaint yourself with the material in the first four chapters before tearing into your set.

EQUIPMENT

While it is not my intention to suggest that beginners equip themselves with a set of professional instruments and tools just to be able to do some simple repairs, there are a few necessary aids needed to undertake such repairs, in addition to the usual simple tools found in every home, such as screwdrivers, pliers, etc.

The first of these aids is a small soldering tool, about 100-watt capacity, preferably one of the instant heating type commonly called a soldering gun. This is not to suggest that you immediately attempt to unsolder and solder parts of your TV set. It is rather a form of insurance in case of *accdients*. For example, if during a simple tube change or adjustment you unintentionally break an interconnecting lead (a number of these seem to be *floating* around the back of every TV set behind the cover), you should be able to repair it without having to run for help. With a soldering gun you need some solder—rosin core type only—never acid core or plumber's type, and preferably of the "thin wire" type. A half-pound spool of this stuff costs a couple of dollars or so in a radio supply house and will last almost a lifetime unless you get into the business full time.

The second piece of equipment that cannot be too highly recommended is a complete set of tubes for your TV set (the picture tube, of course, not included). In hybrid sets (some transistors, some tubes) this advice still applies as far as the tubes in the set are concerned. The transistors are considered later as parts of board assemblies. A set of tubes involves some cost, of course, but actually it involves no additional expense, since defective tubes have to be replaced anyway. The great advantage in having a complete set of spare tubes is most obvious if you have a transformerless set (which are fairly popular) as we shall see presently, but it is no less convenient with transformer-operated receivers. Since the small tubes in a TV receiver will probably have to be replaced at least once during the life of the set, having a set of spares on hand actually involves no additional cost at all.

The third item of great help to any do-it-yourselfer is a schematic diagram or set of diagrams and other service data for your set. This can sometimes be obtained by writing to the manufacturer, giving the model of the set and the chassis number. This information is always printed or stamped on the back cover of the set or the back apron of the chassis. Sometimes the chassis identification is also given on the tube-transistor layout chart usually found inside the cabinet.

There is another and much more readily available source of service information. Instead of writing to the manufacturer schematics, layouts, adjustment instructions and other helpful hints can be bought in almost any radio supply house. While most users of this book may not be very familiar with schematic diagrams, these nevertheless are very useful because they give tube and transistor identification numbers, their functions as well as their relative physical locations in the chassis.

The last item of equipment is a substitute electric cord and plug, known as a *cheater cord*. The name is appropriate, since it enables its user to cheat or defeat the protective system designed into nearly all TV sets, namely prevention of the uninformed person from reaching into a potentially dangerous TV set with the power on. Let me add, however, that it is quite safe and permissible if you observe reasonable precaution. It is my intention to properly inform you regarding the dangers lurking behind that cover.

The safety interlock in the vast majority of TV sets consists of the following. The AC line cord is physically tethered to the protective back cover of the set. When the cover is removed the AC cord is automatically pulled out of a 2-pin male plug on the TV chassis. To operate the set with the back removed requires a line cord identical to the one attached to the cover. This is the cheater cord, and it's available from radio supply houses for a dollar or so.

In a number of recent TV sets the cord is not fastened to the back cover. Instead, it can be unplugged from the back of the set before the back is removed. If this is not done first the cord will automatically disconnect from the chassis when the back cover is taken off. In either case removing the cover also disconnects the cord, and it takes a deliberate action to plug the cord back into the chassis without the cover. This still insures that the TV set will not be accidentally exposed and operating; however, in this case, a separate cheater cord is not required.

Before proceeding with actual troubleshooting procedures it is necessary to outline and explain a very useful and very simple procedure which is applicable to a large majority of TV repairs. This is the tube substitution method of TV troubleshooting. Earlier it was indicated how to carry out tube substitution the hard way—remove a suspected tube from its socket to check it on a tube tester. Of course, if you don't have a tube tester, that means going down to the neighborhood store displaying a do-it-yourself tube tester, checking the tube and buying a new one if necessary. If you have a complete set of tubes for your TV set you can troubleshoot and repair-by-substitution much more easily and quickly.

Parallel Tube Filaments

Figure 5-1 is a typical drawing of a parallel tube hookup. Only the heaters or filaments (the elements that glow when the set is

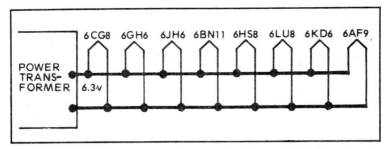

Fig. 5-1. Diagram of tube filaments or heaters wired in parallel. The heavy lines in effect supply each tube individually, thus each heater operates independent of the others. Notice that all tube numbers begin with the same number (the number above the symbol used to represent a tube filament).

switched on) are shown. Notice that every one of the tubes is connected independently to the source of current, most commonly 6.3 volts AC. An incidental clue to such a parallel tube hookup is the fact that the tube numbers (except the picture tube) all begin with the same number, usually 6 (such as 6AH8, 6GH8, 6BG6, etc.), but sometimes 12 (for example, 12AH6, etc.).

Since each tube is connected to the source of the current, without depending on any other tubes to complete the path it becomes obvious that pulling any tube out of its socket will not prevent any of the remaining tubes from glowing or lighting; thus, when a particular tube is suspected, it can be pulled out, a new one of the same type plugged in, and the suspicion confirmed or disproved. In case of a burnout a tube fails to light and can easily be spotted. In certain tubes a positive visual observation or absence of light is not feasible, either due to the coating (the coating is a by-product of its being a vacuum tube and has no other significance) inside the glass, or because of some physical obstruction such as a metal sleeve (called a tube shield). But by carefully touching each tube with your bare hand you can determine its condition by the warmth (sometimes a very high temperature) or coolness to the touch. By contrast, none of these tell-tale clues are available with a transformerless, series-string tube hookup. Other measures, therefore, are necessary.

Series Tube Filament

Figure 5-2 is a typical wiring diagram of heaters in a series hookup. Observe that the tubes follow each other, chain-link fashion. This is called series wiring (like the older or inexpensive type

Christmas tree lights). As in the case of parallel hookup just described, the tube numbering is a reliable clue. In a series hookup there will be quite an assortment of tube numbers, such as 3AH8, 18BQ7, 14SH6, 12GD9, etc. Whenever any of these tubes is removed from its socket, or whenever one burns out, the circuit is opened and all tubes go out. There is absolutely no indication which of the tubes is the culprit, or for that matter, whether more than one is involved. While modern tube burnouts are not too frequent, this does not help when a burnout does occur.

Incidentally, it is the nature of filament circuits that burnouts are more likely to occur in a series-wired set where detection is more difficult. In the absence of a spare set of tubes the only practical procedure is to remove all tubes and check them in a store tube tester. With spare tubes on hand, one-at-a-time replacement can be made until the offender is found. Of course, it is just possible that the very last tube so tested is the defective one, but it is equally possible that the first, or second, or third tube pulled out is the culprit. The advantage of spares on hand is obvious.

In connection with series-wired tubes it must be mentioned that there is one other possibility where all the tubes fail to light when none of them is burned out. This will happen when a protective device, called a fusible resistor or surge protector, burns out. As was mentioned earlier tubes in a series-string circuit are more prone to burnout and shortened life than the same types of tubes in parallel-wired or transformer-type TV sets. To reduce this failure hazard, many manufacturers have incorporated in series-wired TV sets a limiting (or surge absorbing) resistor, wired in as if it were one additional tube in the string; however, since it is not as obvious as a tube and often much less accessible, it can be overlooked.

Fig. 5-2. When tube heaters are wired in series, as they are in this drawing, they depend on each other for power. If one goes out, the circuit is broken and they all go out.

While some beginners may consider this task beyond either their present ability or ambition, others may be quite up to replacing the resistor themselves; therefore, here is a simple procedure for locating and replacing this resistor. Incidentally, this so-called fusible resistor is sold by radio supply houses, often in quantities of three or four in a package, for about a dollar or so. It is not a very critical part, and no precise TV set make and model need be known to buy one of these; a number of different makes and models use the same part. Figure 5-3 shows a simple diagram of the wiring from the AC line cord to the on-off switch to the surge resistor and tubes. This resistor is almost always located immediately after the on-off switch and in most sets is located in plain view and reasonably accessible.

In order to determine whether or not the resistor is burned out, disconnect the receiver from the wall outlet and remove the rear cover. Should it be required to remove the chassis from the cabinet (this is usually not the case), the picture tube and the speaker are unplugged, leaving the chassis free. Most service data clearly identifies what connectors exist between the chassis and the speaker and picture tube. Usually, there are two separate wires from the chassis to the speaker, each with a quick-disconnect device. Sometimes these two wires may interconnect through a 2-pin plug. The picture tube usually has a multipin plug for the same purpose, and in some color TV sets, two such plugs. In addition, there is a high-voltage lead going to a snap button on the picture tube glass. This also must be disconnected, preferably a few minutes after the set has been switched off to allow time for the residual high-voltage to dissipate. Otherwise, an unpleasant jolt may be in store for the bare hand touching this lead. Incidentally, a large number of TV sets have a quick-disconnect on this lead at the chassis end, making it so much easier to handle. This applies to those receivers in which the picture tube is mounted to the cabinet. In some sets, however, especially some of the portable type, the picture tube is mounted on the chassis, and no disconnecting of any picture tube wires or plugs is required.

To determine whether the surge resistor is burned out, the professional TV servicer makes a continuity check with an ohmmeter; however, it is perfectly safe to *bridge* this resistor for test purposes (connecting the two ends together, as shown in Fig. 5-3), then observe whether or not the tubes light. If they do light, the resistor is burned out and should be replaced. If they do not light one of them is burned out.

Series-Parallel Tube Filaments

Before concluding our discussion of tube replacement in series strings we should consider special series-parallel filaments found particularly in some older receivers. So far, to avoid complicating things, I have not mentioned the fact that a series strings (as in Fig. 5-3) the same current flows through all the tubes and that the tubes are purposely designed and chosen so that all tubes in the string, regardless of tube number, require the same current. In contrast, each tube in a parallel-filament TV set may (and sometimes does) carry a different current than its neighbor; they are independent of each other in this respect. Not so in the series string arrangement.

Let us again look at Fig. 5-3. When the string is complete, that is, all tubes light, the current is continuous and the same through the whole string. If, for example, tube V4 (shown dashed in Fig. 5-3) should burn out, all tubes will be out; however, consider a series string circuit of the type found in some of the older sets. Figure 5-4 shows a simplified version for the sake of explanation. Notice that V4 and V5 are, so to speak, together. Should one of these two tubes (say V4) burn out, all tubes will continue to light, and the second tube of the pair (V5) in this case will be overloaded and most likely will

Fig. 5-3. A typical series filament string circuit. Three requirements must be met for all tubes to light. The on-off switch (shown open) must be closed; all the tubes must be good (V4 is shown burned out); and the surge resistor must be good. A piece of wire momentarily placed across the resistor will tell you if it's burned out because the tubes will light when the wire is connected.

burn out or be permanently damaged. This type of hybrid tube hookup was employed by some manufacturers because there was not available a full choice of tubes for a simple series hookup as in Fig. 5-3.

In troubleshooting a TV set of this type it is necessary to have the diagrams (mentioned earlier) showing the tube hookup. Before removing any tube for possible replacement it is but necessary to look at the tube heater hookup, which is always shown in the diagram. Any tubes connected like V1, V2, V3, V6, V7, or V9 in Fig. 5-4 may be removed, tested and replaced without any fear of damage. Any tube hooked up in pairs, like V4 and V5, should be removed in pairs; that is, the set switched off, both tubes removed, each tube replaced in turn, then the set switched on to observe the effect of the substitution.

A final suggestion on hybrid tube wiring: in some (very few) obsolete sets, a tube or two may be paired not with another tube, as are V4 and V5 in Fig. 5-4, but with a heater element (called a resistor). The circuit might look like V8 in Fig. 5-4. Observe that should V4, V5, or V8 burn out or be removed from their sockets, the remaining tubes will still be able to light, but the heater element marked R or the second tube of the pair will be carrying all the current and may be damaged as a result. In such cases the safe and proper procedure is to switch off the set, remove the tube or tubes, substitute others, then switch on the receiver and observe the effect of the tube replacement.

TUBE PULLING

Since it is assumed that you have no experience in even the relatively simple task of tube changing the following instructions should further simplify this task. In addition the precautions given here will, as a minimum, save time and prevent damage. More important they may prevent potential (minor) injury as well as discouragement.

Because of the prevalent practice to cram the maximum amount of electronics into the minimum possible space, some tube locations are often less than ideal for removal or replacement. This requires extra care in removal and replacement of tube shields and tubes. The general purpose of this admonition is twofold. First, the removal of a tube must not affect or damage any adjacent part. Delicate components, fragile leads and potentially shock-producing (exposed) voltages may lurk in the immediate vicinity of the tubes. Plain care and

Fig. 5-4. Some older TV sets have a series-parallel filament circuit, where two tubes are (V4 and V5) connected in parallel or a resistor is placed across a tube filament.

common sense are all that is required to insure that when the tube shield and tube are handled, other components in the vicinity are left undisturbed. Second, proper care must be exercised both in removal and replacement of tubes so that tube pins are not bent or broken, sockets are not damaged, or, what is less obvious, that socket connections are not loosened by careless or forceful handling. This is easier done in some cases than in others, as I shall show presently.

Octal Tubes

Octal tubes generally are known as the older generation. They are glass or even metal (rare) tubes with bakelite bases. Figure 5-5 shows the base and matching socket of an octal tube. Although they are called octal tubes, they may not have all eight pins. But whether the tubes have five, six, seven, or eight pins, they are all laid out on a symmetrical pattern of eight around a circle. The octal socket, by contrast, and for reasons of simplicity in manufacturing and stocking, has all eight pinholes, with connecting lugs underneath. Any unused lugs (as when a five-pin tube is to be plugged in) are simply left blank or used for anchoring or supporting other electronic parts in the vicinity.

Notice that the tube has a key (corresponding to the key on motor shafts and pulleys) and the socket has a keyway which must be positioned in one correct way only, otherwise the tube will not readily plug into the socket. Here is a good example of what is meant by plain care and common sense during tube removal and replacement. It is quite possible, with a little force, to wear damage the fiber or phenolic keyway in the socket and finally "get the tubes in" the wrong way. The proper procedure, and this can actually be done in the dark, that is, without seeing the keyway, is to hold the tube upright over the center of the socket. Then, and without any downward pushing whatsoever, the tube is moved slightly until just the tip of the rounded center pin barely enters the center hole of the socket. Now, slowly turn the tube until it drops a bit into the socket. (Up to now, only the rounded end of the tube center pin was pivoting in the hole.) When the tube drops all the pins will begin to engage the corresponding socket holes. Gentle but firm downward pressure will fully seat the tube. An experimental trial with a tube and an accessible socket takes less time than it takes to describe it.

To remove the tube a straight upward pull is used, taking care that your hand (with the tube in it) doesn't come up suddenly and hit something inside the TV set. Incidentally, a straight, perfectly vertical pull upward is not always easy, but with caution and judgment it is permissible, while pulling the tube up, to wiggle it slightly from side to side. The tube pins are strong enough and the socket pins flexible enough to permit doing this without causing any damage.

Tube Shield Handling

A metal sleeve over the tube, where used, is an essential part of the circuit; it is not ornamental or for mechanical protection. It is absolutely imperative that the shield be replaced and properly seated in its retaining base or clips after the tube has been replaced. Failing to replace a shield may not produce an immediate and obvious malfunction, but many a complaint of poor TV performance has been traced to the absence of shields as the result of failing to replace them after tube changes. In many TV sets the shields on the tuner are soldered in place and are not completely removable. They are, however, collapsible to half of their normal height (by simply pushing down on the rim) so that the tube can be removed. After tube replacement the shield is returned again to its original height.

All-Glass Tubes

These have been generally classified as miniature tubes and included, up to a few years ago, seven-pin and nine-pin tubes of rather small size. Since then, however, a new group of tubes in-

Fig. 5-5. Sketch of an octal tube and socket.

cludes types which often are as large as the largest base-type tubes. The method of handling is the same for all types so they are discussed as a group.

Seven-Pin Miniatures—Pins of these tubes are made of relatively soft metal and will withstand a limited amount of bending and straightening; however, if any pins are bent in the process of plugging into the socket it can cause poor contact and resultant performance troubles. Figure 5-6 shows a bottom view of a seven-pin miniature tube and its socket. Observe that unlike octal tubes, there is no key, no keyway, and no mechanical guide for correct insertion. The seven pins are symmetrically spaced on an eight-pin pattern, with one pin omitted. Therefore, more care is required in removal and especially in plugging in such a tube. In the latter case visual observation of the tube and socket is essential, especially the first time. As in the case of octal tubes it is advisable to experiment with a tube that is both accessible and visible. The proper procedure for plugging in such a tube is as follows:

Align the tube so that the blank space on the tube faces the blank space of the socket. In other words, pins 1 and 7 (wide spacing) face holes 1 and 7 on the socket.

With the tube resting on the socket (of its own weight), rotate the tube very slightly back and forth to get all seven pins to the point

Fig. 5-6. Drawing of a seven-pin miniature all-glass tube and its socket.

where they just begin to find the corresponding socket holes. This is best done by feeling rather than looking.

Firmly push the tube straight down. If the pins are aligned the tube will move downward, although it will offer some resistance.

Continue the downward pressure until the tube is fully seated. This, too, is done by feel, although it may also be seen.

If the tube does not seem to begin moving downward when you first apply pressure remove it and carefully examine the pins. Due to possible improper alignment, one or more of the pins might have pushed against the socket material and become bent. Carefully straighten the pin and repeat the process.

Now and then a gadget appears on the market which is intended to assist in plugging in these tubes. Most servicers, however, do not use them.

As mentioned in the case of the octal tubes it is a good idea to practice removing and replacing a tube on a fully accessible socket where you can see exactly what happens. What has been said about tube shields before applies even more so here since most of the critical and sensitive functions in modern TV sets are performed by miniature tubes and the shield is an important part of the circuit.

Nine-Pin Miniatures—Figure 5-7 shows the tube pin layout and socket of the nine-pin miniature tube type. What has been said about the seven-pin tube and socket applied equally here. Nine-pin tubes are somewhat larger in diameter, too.

Other All-Glass Tubes—There are other types of all-glass tubes in use in TV sets, but these are basically no different than the

Fig. 5-7. The nine-pin miniature tube and socket are similar to seven-pin tubes.

two types just described. For example, a series of tubes known as Compactrons are considerably larger than the miniature types just described; in fact they are among some of the largest tubes found in a modern TV set. Their base structure differs somewhat from the miniatures, and a word of caution with regard to this difference is in order. All such tubes are, of course, evacuated the same as other types. But miniature tubes previously discusses have the bulb tip at the top, while Compactron types have this tip on the bottom right in the center of the pin circle as shown in Figure 5-8.

While handling this type of tube care must be exercised that the glass tip is not accidentally broken. As shown in Figure 5-8 the

Fig. 5-8. Some receivers use larger all-glass tubes like this sketch of a Compactron and its socket.

socket has a clearance hole for the tip. Nevertheless, it is possible to strike it against some object on the chassis. Breaking this tip will not cause any violent implosion but it will certainly destroy the tube.

Tubes With Top Caps

There are at least two tubes in a modern TV set which handle very high-voltage (in the thousands). Because of this one connection to such a tube is brought out to a metallic button (top cap) on top of the tube instead of to one of the base pins. Furthermore, in the interest of safety, the connecting top cap lead is often enclosed in a nonmetallic (bakelite, phenolic, etc.) outer shell. While the top cap button on the tube is electrically soldered to the appropriate internal tube element, the button is only cemented to the glass; quite often the cement separates from the glass, leaving only the soldered wire to hold the button. While this does not call for discarding the tube, it calls for special care in removal and replacement of the top cap lead.

To remove a connection from a loose button first try (gently) to rotate the cap a bit. If it responds gradually pull it straight upward while rotating it a bit. Under normal conditions, such rotation will be accompanied by a very slight scraping of metal against metal, indicating that the button on the tube is rigid while the cap is rubbing against the button. Since there is motion between the two metal surfaces, it is safe to pull the cap off; however, should this slight rotation be smooth and noiseless, it probably (even if not 100% of the time) means that both the cap and button are turning together because the button is no longer cemented to the glass. Further twisting is likely to break the connection to the tube element and thereby destroy the tube. In such a case it should be possible, using both hands or perhaps a thin screwdriver blade, to hold the loose button on top of the tube while working the cap loose. If a subsequent test in a tube tester shows the tube to be in operating condition the looseness of the cap may be ignored. If desired any household cement may be used to attach the cap before plugging it back into the TV set.

TESTING BY SUBSTITUTION

Now that we have become acquainted with the mechanics of tube pulling and replacement, we can proceed to troubleshoot by tube substitution.

Remove the back cover with the line cord assembly or the line cord and the cover separately if they are not attached to each other. Connect the set to the AC wall outlet, using the cheater cord or the TV set cord if it is free.

Switch the TV set on. Observe as far as possible whether any tube fails to light. In case of the transformerless TV set all or none will light. I shall outline the procedure for this type later. For a parallel-wired (transformer type) set only the burned out tube will not light. All others will light.

If it is difficult to see whether or not a tube lights determine by the touch method which tube if any does not feel warm. If all tubes light or are warm to the touch the defect is not due to a burnout, although a tube may still be defective. Be aware that some tubes operate at very high temperatures, sufficient to cause painful burns.

Determine from the troubleshooting procedures that follow which of the tubes may be involved. Remove the tube, following the procedures and caution notes. Substitute a good tube for the suspected one. In cases where more than one tube may be the cause, change only one tube at a time.

Switch on the TV set, allow a few minutes for warmup and stabilization, then observe the effect. If the tube was at fault the set should now perform normally. If it does not return to the section in the book describing the malfunction and continue the procedures given there.

Chapter 6
Operating Adjustments

As we continue to explore TV troubleshooting the next logical step is a knowledge of the operating controls—not only knobs on the front of the receiver but those mysterious looking adjustments in the back.

VERTICAL SIZE

The normal TV set has sufficient vertical size, or as it is commonly called *height*, to fill the screen completely with quite a little to spare. In other words, the back-of-the-set adjustment called vertical size (V-size) or height could make the picture as much as 30% higher than the screen height if turned far enough; however, normal design provides for a normal size picture with the control in approximately its midway position.

Another important consideration before adjusting the vertical size or height control is the avoidance of distortion. The normal picture format (or aspect ratio) by technical and NTSC standard is four units wide by three units high; that is, for every four inches of picture width there is a corresponding three inches of picture height. Under these conditions a circular emblem will look like a perfect circle, not egg shaped. In adjusting the vertical size or height care should be taken that the picture is not stretched to the point of distortion. This is best done with some circular object on the screen. Some stations still show a circular test pattern a few minutes before going on the air; however, it is not necessary to wait for a test pattern—many emblems and other commercial symbols shown between programs are circular and will serve for vertical size adjustments. Possible alternatives for height adjustments, therefore, are as follows:

- If the picture height is slightly below normal, adjust the vertical size or height control until the picture just covers the screen height. Check for nondistortion by observing the circularity of a suitable object on the screen, preferably a large one for ease of correct adjustments. For all adjustments where observation of the screen is necessary, a mirror is essential. The professional TV servicer has a special mirror on a stand for the purpose, but any mirror propped up in front of the screen so it can be viewed from behind the set is quite satisfactory.
- If the control setting for full screen height is at or near its extreme position (little rotation remaining) the vertical amplifier stage is not operating properly. While it is just possible that a defective resistor, capacitor, or transformer, or even a voltage change, is responsible for the loss of height, the most likely and most frequent cause is a weak tube in the vertical circuit.

Referring to the typical functional block diagram in Chapter 2 we notice that there is a vertical oscillator tube and a vertical amplifier tube in every TV set. Sometimes these functions have different names for the same functions; thus vertical multivibrator is a synonym for oscillator.

V-output is quite commonly used for a shortened name of the vertical amplifier. What is significant is the fact that in the vast majority of modern TV sets, these two functions, vertical oscillator and vertical amplifier, are on one envelope and can be located easily by observing the number description, such as V6A and V6B, meaning parts A and B of tube V6. After locating and removing this two-in-one tube from its socket, if the TV set is switched on it should have normal sound and a simple bright horizontal line across the center of the picture tube. If, on the basis of probability, the pulled tube checks (in a tube tester) as *weak* or *poor*, replacement with a new tube will solve the problem but will almost always require adjustment of the *V-size* and perhaps also the *V-line* (linearity) control. The procedure is as follows:

1. Plug in the new tube and switch the TV set on. Wait a few minutes for warmup.
2. With a mirror positioned so you can see the picture from your position behind the set, reduce the V-size until the picture is too small to cover the screen both at the top and bottom; that is, until some blank screen shows.
3. Wait for a circular pattern or emblem to appear on the screen and observe its symmetry. It is quite easy to tell whether or not a circular shape is 100% symmetrical or not. Figure 6-1 is an example of patterns you may encounter during this adjustment.
4. If the circular pattern resembles Fig. 6-1A, that is, if two halves are symmetrical, adjust the V-size to increase the pattern or display until all four quarters are as identical as possible, or until the circle is no longer flattened at the top and bottom. This will produce the closest approach to a perfect circle. (There may be some minor irregularity due to horizontal and other imperfections, preventing the attainment of a 100% perfect circle, but if this deviation is very slight, no furher correction is necessary.) The pattern should look pretty much like that in Fig. 6-1D.

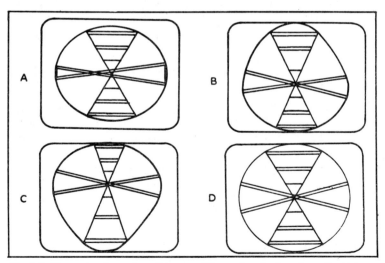

Fig. 6-1. These drawings show three conditions of vertical nonlinearity. Using a circular emblem or insignia adjust the vertical linearity for a symmetrical circle as indicated at D. The circle at A is flat on both top and bottom; at B the top half is stretched and the lower half compressed. The nonlinearity at C is the exact opposite of the condition at B.

5. If the final, most nearly circular pattern is obtained before the screen is fully covered in the vertical direction or, conversely, if you have to turn the V-size control to the point where the picture runs beyond the top and bottom edges of the screen in order to get a circle, then the width (horizontal size) adjustment is incorrect and will have to be corrected before completion of the vertical adjustment; however, at the moment we are proceeding on the assumption that only the vertical circuits require adjustment. (We shall return to the horizontal adjustment presently.)

VERTICAL LINEARITY

If the reduced image on the screen resembles Fig. 6-1B or 6-1C, a second, closely related vertical adjustment must also be made. This control varies the vertical picture symmetry, usually called vertical linearity. While it is difficult to tell a nearly perfect shape by viewing an action scene on the screen, such is not the case for distorted proportion. For example, Fig. 6-1B corresponds to excessive stretching of the top of the picture; people seem to have very short legs and their heads seem to come to a point. The opposite extreme as in Fig. 6-1C produces long-legged, high-waisted people with very low foreheads. Correction of either requires the adjustment of the V-lin (vertical linearity) control, again using a circular pattern of some kind, plus the aid of a ruler and a grease pencil, as follows:

1. Mark a horizontal line across the screen, midway between the top and bottom; use a ruler for a fairly accurate division of the height.
2. Adjust the V-size control, as previously, for a circle smaller than the full screen height, exposing blank space at the top and bottom.
3. Adjust the V-lin control until the upper and lower portions (above and below the painted line) are as nearly of the same height as you can tell with the naked eye (measurement is not necessary).

VERTICAL ROLL

In a normally functioning TV receiver the picture will at times slide up or down one frame. This may happen when the set is first switched on, when channels are changed or when programs are switched at the transmitter. At all other times the picture actually does not "want" to slip or roll as it is controlled by and synchronized with a signal from the transmitter (vertical sync, see Chapter 2). If however, a TV set exhibits frequent rolling, requiring repeated adjustment of the V-hold control, the fault is most likely in the sync circuits.

An examination of the typical block diagrams, and especially the diagram of the particular set in question, will reveal that there are such functions as sync amplifier, sync separator and sync clipper, or perhaps combinations of these functions, performed by one or more tubes all having the common term sync. A defect in one of these tubes is the most likely suspect causing vertical roll. As shown in Figs. 6-2, 6-3, 6-4 and 6-5, these are dual- or triple-function tubes, identified as sections A, B, or C (V5A, V5B, etc.). Each section of such tubes may be involved in separate unrelated functions so that removal of one tube actually disables two or three separate functions; thus, in Fig. 6-2, one-half of V8 (V8A) is a sync clipper amplifier, while the other half (V8B) is the vertical oscillator. With V8 out of its socket the picture will collapse to a thin horizontal line.

In most cases a sync tube serves both vertical and horizontal sweep circuits. Removal of this tube will, therefore, disable both vertical and horizontal sync, and the picture will roll vertically and tear (zig-zag fashion) horizontally. As soon as a good tube is replaced in the socket, everything should return to normal, provided the tube was at fault.

Other Causes of Sync Failure

Weak or defective sync tubes are not the only potential causes for vertical roll. Referring again to the block diagrams it will be noted that the signal to the sync amplifier. Very often this stage, not the following sync circuits, is responsible. Should the video amplifier be either very poor or outright defective, a very poor sync signal will leave this tube; therefore, the input to the sync stages is inadequate.

Of course, in such a case the picture quality will also suffer, most often exhibiting a lack of highlights and contrast. While advancing the contrast control may result in some improvment to the picture, this adjustment may actually be detrimental to normal sync output; thus, chronic picture rolling in a set having a weak picture may well be due to a subnormal video amplifier. Also, while a poor picture may be due to one or more substandard tubes anywhere from the antenna to the video amplifier, it is rather easy to tell which portion to the set is responsible. A weak "thin" picture that is otherwise clean is almost certain to be due to a poor video amplifier, while substandard functioning in the TV tuner or in the IF amplifier is almost certain to produce a snowy, perhaps wavy, picture.

FREAK SYNC TROUBLE

There is one other potential case of sync problems worth mentioning here—interference from another TV station on the same channel. Standard FCC channel allocations are such that under normal conditions the geographic separation between stations operating on the same channel preclude any likelihood of the distant station interfering with a local station on the same channel. Modern TV sets have, in addition, a certain amount of built-in immunity from this possible interference because they tend to favor the much stronger local station; however, during some not-too-frequent freakish atmosphere conditions, when distant signals tend to come in rather strongly, interference may result. This is due to the fact that although technical standards (frequencies, sync, sweeps, etc.) are the same throughout the United States, very minor and quite tolerable differences between those standards will cause severe interference with the local station sync and, consequently, with the picture as a whole; however, since these are very transient conditions, occurring on some stations only, nothing should be done about it. It is described here merely to acquaint you with the phenomenon, and caution you not to rush into making adjustments.

HORIZONTAL SIZE OR WIDTH

Symptons of insufficient width are quite obvious: the picture does not cover the full width of the screen and, quite likely, has shrunk more on one side than on the other. As in the case of vertical size, either tube deterioration or adjustments, or (most likely) both, are responsible. Referring to Fig. 6-2 we see that tubes V10A, V10B, and V11 are involved. In Fig. 6-3 V9 and V10 are involved. In Fig. 6-4 V10 and V11 are involved, while in Fig. 6-5 V9, V10A and V10B are involved.

When insufficient width is the only complaint, the horizontal oscillator function block may not be involved; however, since adjustments are required after substituting a higher performing new tube for a worn out one, the horizontal oscillator block may also be involved since some of the adjustments are located there. Again referring to the typical block diagrams, other tubes and functions are shown, because they are intimately related to the horizontal amplifier.

In troubleshooting for insufficient width first suspect the horizontal amplifier. Proceed as follows:

Fig. 6-2. Block diagram of a typical Black & White TV set with parallel-wired tube filaments.

101

1. Switch off the receiver. Remove the back cover and AC cord. Use a cheater cord if necessary.
2. Carefully remove the top cap from the horizontal amplifier tube and remove the tube. This cap may be very tight or sometimes frozen. Observe the suggestions given in Chapter 5 for removing the cap without damaging the tube.
3. Install a new tube and replace the top cap securely. This is a high-voltage point; a loose cap may cause arcing and even damage to the horizontal output transformer. If this tube is located inside a metal enclosure or cage be sure to secure the enclosure after tube replacement. Practically all of the components in this cage operate at extremely high voltages.
4. If the horizontal amplifier tube replacement does not correct the insufficient width one or two other tubes may be at fault. Looking again at the typical block diagrams, Figs. 6-2 to 6-5, we notice that each has a tube called a damper. Without going into the technicalities of the function of the damper, it is sufficient for our purpose to know that this tube contributes in large measure to the normal operation of the vertical and horizontal amplifiers by providing them with a voltage boost. Should this tube be below par (and this does happen frequently) the horizontal amplifier will not provide sufficient output for full picture width. Again, caution must be exercised here since the damper tube often is located inside the high-voltage cage. Another tube that may be responsible for insufficient width, also by virtue of providing insufficient voltage to the horizontal amplifier (as well as to all other tubes in the set) is the rectifier tube, where one is used. More and more TV sets nowadays no longer use vacuum tubes for rectifiers; semiconductor diodes, usually marked D (D1 and D2) on the diagram, are preferred for many reasons, not the least of which is a long, trouble-free life. In the previously mentioned diagrams the rectifiers are marked V14 or Z *diodes*. When a tube-type rectifier is used, and if it is weak, the symptoms will be more than just insufficient width. The picture height may be subnormal, sound may be below par, picture may be dim and flat, etc. Replacement of semiconductor rectifiers is in my opinion beyond the scope of a beginner and is not recommended.

Fig. 6-3. Block diagram of another popular Black & White TV set.

5. If it is obvious that the picture width is correct no further adjustments are required; however, this is the proper time to adjust the set if, as is often the case, the picture is too wide, wrapping itself partly around the tube, causing cutoff of left and right edges. This can best be seen when some printing or writing is shown, such as at the opening of a motion picture. If it is apparent that the picture is too wide, proceed to steps 6 and 7.
6. If you have a set of diagrams or service information for the set examine it to find what type of width adjustment is provided. If no such data is at hand proceed to step 7.
7. Examine the back of the TV set with the cover removed. Compare it with the typical diagrams in Figs. 6-2 through 6-5. Notice that on these diagrams three different kinds of width adjustments are indicated: pot, slug, and points. While all of these are meant to achieve the same purpose the procedure for each is different and must be understood before attempting to use any of them.

Pot—This adjustment is of the volume control type and may be a knurled, round shaft like that inside the knobs of the ordinary radio, or it may be a slotted shaft designed for screwdriver adjustment. In either case, it is an adjustable potentiometer (pot for short) which can be rotated less than one full turn. It is the simplest to adjust and can easily be reset to its original (before adjustment) position. It is worth mentioning a second time that it is wise for the beginner to always observe and, where possible, record the position of an adjustment before changing it. In this way you can always go back to the original condition should you feel the adjustment makes things worse or is ineffective.

With a pot type of control slowly rotate the shaft (using a mirror to watch the picture) until the width is less than full screen. If the picture extends more or less equally on both left and right edges, rotate the shaft in the opposite direction until the picture just barely extends beyond the edges of the screen. About ¼- to ½-inch is all that should be *wrapped* on each side. The correct width setting is that which does not mask any part of a line of printed matter (the list of actors, etc., displayed at the beginning of a movie is most suitable).

Slug—A slug-type width adjustment is found in many TV receivers and requires a different technique, as well as some precautionary measures. The word slug refers to a carbon-like rod that can be made to slide in or out of a fiber sleeve on which a spool of wire

Fig. 6-4. Block diagram of a Black & White TV set with series-wired filaments.

is wound. The whole assembly is called a *width coil* or (a bit more technically) a slug-tuned coil. The tuning or adjustment of width is made by inserting the blade of a special tool (or midget screwdriver) into the protruding slot of a threaded screw-like shaft. Figure 6-6 is a simple sketch of such a coil.

In contrast to the just described pot adjustment the slug can be turned as many as eight or more times. In fact there is no stop on this adjustment, hence the precaution: Do not keep on turning a slug more than three or four turns in either direction. To do so may unscrew the slug completely, letting it fall out or, worse, fall into the dark recess of the TV chassis. Whichever way it falls, there follows a messy job of replacing it, certainly one to be avoided by the beginner. Fortunately, TV design is such that this slug is normally in an approximately midposition, allowing a minimum travel of five or six turns in or out—certainly enough for any width adjustment in the field. Normally most width coils are designed so that turning the slug out (counterclockwise from the screwdriver end) will increase the width, and vice versa. The procedure, therefore, is the same as the pot shaft rotation, except that here a few full turns may be required. The correct procedure for adjustment is as follows:

Rotate the slug clockwise until both left and right edges of the picture are visible; the picture is now too narrow. Rotate the slug in the other direction until a slight overlap or *wrap* is achieved on each side. Of course, too much overlap cuts off the picture.

Points—In this method of width adjustment a movable voltage lead can be attached to various voltage *points*. By changing the voltage applied to a portion of the horizontal amplifier the width of the picture can be varied slightly. Usually there are three or four tie points to select. In somes cases not connecting the movable jumper lead to any voltage point provides a possible setting for the horizontal width. Since the movable jumper is usually insulated this procedure can be made with the TV under power.

Other Width Adjustments—In some TV receivers, a mechanical (magnetic) adjustment is provided for picture width control. No two of these are alike, and adjustment should not be attempted without some sort of service instructions. Two of the most common mechanical adjustments are the width sleeve and the width tabs.

The width sleeve is just that—a metallic sleeve on the rear of the neck of the picture tube. This sleeve is easily adjusted by a combined sliding-rotation motion (it may sometimes stick to the

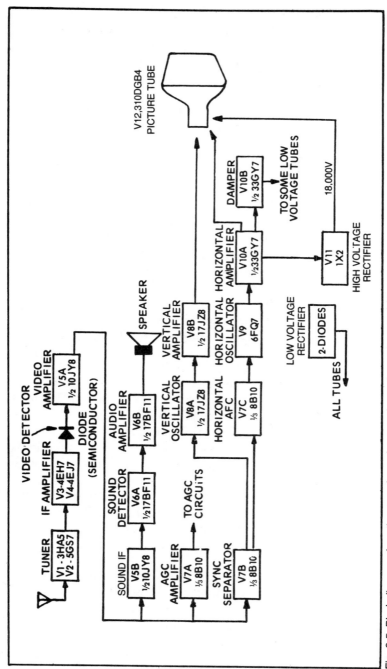

Fig. 6-5. Block diagram of another transformerless receiver using series-wired filaments.

glass until dislodged) until the correct width adjustment is obtained. Again I repeat the need for caution: there is no danger from either heat or high voltage in this procedure, but neither is there room for carelessness. Since the sleeve sliding along the neck of the picture tube also affects horizontal linearity, we shall return to this adjustment presently.

Width tabs function in much the same manner as the sleeve. The tabs have the same effect on width as a sliding sleeve. Since there can be a great variety of mechanical means of varying picture width, no standard, uniform procedure can be given, except as described above.

HORIZONTAL LINEARITY

As mentioned previously in connection with the procedure for horizontal width adjustment and correction, the horizontal linearity adjustment is most conveniently made at the same time. In some sets the two interact so that it is almost mandatory to do them together, unless a nonlinearity problem exists without any accompanying reduction in width. In that case it may be required to reduce the width in order to be able to see the nonlinearity.

Horizontal linearity can best be understood by referring to the discussion of vertical linearity. In simplest terms horizontal linearity means that the left and right halves of the picture are symmetrical. More specifically, and this applies to the vast majority of TV receivers, horizontal linearity means that a circle does not appear like an egg, and that the right side is neither stretched or compressed. This is not always obvious when the picture is wrapped around the tube sides; therefore, nonlinearity is best shown when the width is reduced to a little less than the edges of the screen. As in the case of the vertical linearity a circular object is most helpful in observing horizontal linearity, although this is not a must. It should also be kept in mind that although the standard aspect ratio (picture width versus height) is four to three—that is, the picture is a rectangle, not a square—a circle will appear (not an egg lying down) on a properly adjusted TV set.

Horizontal linearity adjustments are not the same on all TV sets; however, most sets have one of three types of adjustments. In earlier TV sets a potentiometer adjustment on the rear apron of the set provides correction for nonlinearity. In other sets a slug adjustment, identical in appearance with the width slug adjustment described earlier, is used. In still other TV sets a metallic sleeve on the

Fig. 6-6. Drawing of a typical slug-tuned coil.

tube neck, by itself or in conjunction with a shaft adjustment, serve the same purpose.

Finally, some TV sets have no adjustment whatsoever, relying on the normal operation of the circuit (which was originally designed for proper linearity) to insure linear horizontal configuration of the picture. It is important in the latter case to make sure that the horizontal oscillator and horizontal amplifier tubes are not significantly below normal. Any tube that tests weak on a tube tester is a likely suspect for causing nonlinearity, even though the tube seems to work in the set. The step by step procedure for linearity correction is:

1. Make sure that the horizontal oscillator and amplifier tubes are good.
2. Reduce the picture width until both edges are visible. If either the linearity or centering (see *Centering*) are incorrect the blank spaces may differ in width. With both edges clearly visible examine the picture carefully to determine whether it is actually nonlinearity or improper centering. If after examining a suitable scene, preferably a circular object, it appears that both edges look symmetrical but the blank spaces are of different width, no linearity adjustment is made; the picture is centered according to the directions that follow. If, however, there is definite distortion on one edge (usually the right side), proceed to step 3.
3. Rotate the linearity adjustment shaft a small amount at a time, first to the right then to the left, and observe the

effects on the picture. Figure 6-7 shows two opposites of nonlinearity. In Fig. 6-7B the picture should be made to shrink toward the center of the screen until symmetry is reestablished. In the case of Fig. 6-7A a certain amount of picture stretching is called for, until the picture most nearly resembles Fig. 6-7C. Most likely this will also affect width.

4. Adjust the width control, as described earlier, until the picture again just begins to wrap around the edges of the tube. It may be necessary to slightly adjust the width control during the linearity adjustment, but a final width touchup may still be required.

5. If the linearity adjustment is a slug type, proceed as for the slug-type width adjustment: in other words rotate the shaft clockwise or counterclockwise two or three turns until the effect is observed to determine if it is going the wrong way or not; then rotate it in the proper direction until the best linearity is achieved. The same precautions given for slug width adjustments apply here. If three or four turns seem to make no difference, discontinue the adjustment; the fault does not lie here. There are a number of components (resistor, capacitor or peaking coil) which, if defective, would cause horizontal nonlinearity, and no amount of adjustment can correct this. A professional servicer is then needed.

6. If the linearity adjustment is mechanical or a similar type (as verified either by the service data or as may be indicated on the interior of the TV set cabinet) the specific procedure should be follwed. One popular method as previously mentioned is to slide the metal sleeve on the neck of the picture tube until the picture width is insufficient to cover the full screen width. Rotate the linearity adjustment shaft on the back apron of the TV chassis until the best linearity is obtained, following the general criteria outlined previously. When linearity has been achieved slide the sleeve on the picture tube neck until the picture fills the screen to a little beyond the edges, observing the effect if any on the linearity. These two adjustments (width and linearity) are somewhat interrelated; so it may be necessary to go through the procedure a second time until optimum linearity with proper width is obtained.

Fig. 6-7. Drawings A and B show typical horizontal nonlinearities in comparison with C. Compression on the right is illustrated at A; at B the picture is stretched on the right.

HORIZONTAL DRIVE

Although the horizontal drive adjustment, found in some TV sets, is neither a width or linearity control it can affect both; therefore, it should not be neglected during width or linearity adjustments. If the suggested procedures for correction of nonlinearity have been exhausted without achieving acceptable linearity, a check of the horizontal drive adjustment should be made. This is invariably a shaft-type adjustment, having a maximum rotation of less than one full turn.

After noting the original setting, just in case, the control is gradually rotated in one direction while observing the effects on picture linearity, as described previously, that is, with the picture not quite reaching the edges of the screen. If the linearity worsens the control should be turned in the opposite direction until effective correction is achieved. It is quite possible that in an attempt to obtain some additional width, the horizontal drive control was advanced to the point of overloading, distortion, and nonlinearity. The correct setting is that which produces the best linearity at normal picture width. The drive may have been advanced to compensate for worn tubes in the horizontal oscillator/discharge/amplifier circuits, but it should be checked and if necessary adjusted when the old tubes are replaced with new ones.

PINCUSHION CORRECTION

During picture linearity adjustments it may be necessary to manipulate the pincushion correction magnets, a part of the picture tube neck assembly in some sets. Pincushion magnets are usually sliding metal tabs or bars and are intended to correct for some inherent design characteristics of the picture deflection system. Looking at the back of the tube, there is a collar-like assembly called a yoke positioned at the far front end of the tube neck (see Fig. 6-8). The function of the yoke is to deflect (it is called a deflection yoke) the picture to a perfect rectangular format. Actually, however, and especially in large-size picture tubes, some distortion of the format, called pincushioning, takes place. The picture seems to be stretched as by pulling on the four corners, leaving the sides somewhat caved in (see Fig. 6-9).

Pincushion tabs or bars are fastened to the yoke structure and may be loosened for sliding, after which they are tightened again. As in other cases of nonlinearity described previously these adjustments can best be made when edges of the picture are visible; hence as part of the other adjustments.

A note of caution is in order here. The tube and deflection yoke assembly are rigidly secured. The pincushion correction was made during installation of the original tube. It is usually not necessary nor advisable to make pincushion adjustments unless there is clear

Fig. 6-8. Cross section of an older Black & White picture tube neck assembly. 1. deflection yoke; 2. wing nut to hold yoke; 3. focus coil found in older sets; 4. focus adjustment; 5. centering tabs; 6. ion trap magnet; 7. tube base.

evidence of distortion or when the picture tube is replaced (by a professional TV servicer, of course).

In some TV sets the pincushion correction is entirely automatic by virtue of a sensing, feedback and compensating circuit. Since this is usually a solid-state circuit and very reliable, there is little likelihood of malfunction during the life of the set. In case of failure, and after making sure that the distortion is not simple vertical or horizontal nonlinearity, replacement of the appropriate module board would be required. It should be noted, however, that even in some of the so-called automatic self-correcting circuits in some recent sets, there is a manual presetting adjustment, somewhat akin to the presetting of the vertical hold, from which the automatic correction circuitry takes over.

CENTERING

Another adjustment related to the various dimensional corrections just discussed is centering. The centering adjustments position the picture so it extends equally in all four directions. When the vertical size is adjusted both the top and bottom edges should be

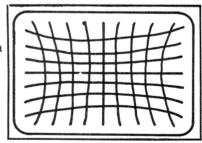

Fig. 6-9. The bowed lines indicate a need for pincushion correction.

reached simultaneously if the picture is centered; likewise, when the width is adjusted both sides should expand at the same time. In very old TV sets, but also in some recent color TV sets, two separate shaft adjustments were used, one marked *V-cent*, the other *H-cent*.

Centering adjustments are very simple; the controls have a total travel of less than 360 degrees (less than one full turn) and can safely be turned back and forth until proper picture centering is achieved. In some TV sets the centering adjustments (if any) are mechanical-magnetic types, centering depending primarily on the original positioning and assembly of the deflection and focus components on the neck of the picture tube and only secondarily on after-the-fact adjustments.

The most popular type of centering consists of a pair of tabs, flat metal, magnetic strips or bars, which can be rotated toward and away from each other until centering is achieved. One tab affects primarily the horizontal positioning, the other the vertical. Again, these are rather interdependent adjustments, and should, therefore, be made simultaneously. A variation of the latter type is the single tab centering device. It can be moved both horizontally and vertically (in fact, it can be made to hove in various directions simply by pushing down on it and sliding it sideways at the same time). Still another centering adjustment is a thumbscrew on the picture tube neck components. This type requires only a clockwise or counterclockwise rotation to achieve centering.

FOCUSING

The focus adjustment varies the sharpness of picture detail. As in the case of the other adjustments, there are different ways of achieving the same end result, and different manufacturers use different methods; however, the procedures given here apply.

- **No Adjustment**—There are a number of TV sets with what is called fixed focus; the picture tube is designed and constructed so that the picture is always in focus, although there may be a provision for correction (by a servicer) in the rare case when this type is out of focus.
- **Focus Thumbscrew**—Some focus controls are in the form of a short, flexible steel cable with a thumbscrew at the end; the cable extends through the back cover of the TV set. Rotating the knurled end of the cable adjusts the sharpness of the picture. Sets with this type hopefully have long since been put to rest, but once in a while you'll see one.

- **Sleeve Adjustment**—This focus correction device is in the form of a ring or sleeve on the neck of the picture tube. Sliding the ring or sleeve slowly along the neck of the tube adjusts the focus of the image. In those TV sets in which the width control is also a sleeve-type adjustment, it will be found that the width adjustment is very much forward on the picture tube neck, while the focus sleeve is very close to the rear end of the picture tube.
- **Shaft Focus Adjustment**—This is a control similar to a volume control on the panel of an ordinary radio receiver. This type of control is found on some (usually older) black-and-white TV sets and on practically all color TV sets where focusing is very much of a requirement.

In all focusing adjustments, it is essential that the picture be watched at close range in order to obtain the sharpest focus possible. While any fine detail of the picture can be used for observing the effect of focusing adjustment, it is best to use the horizontal lines, preferably on a bright portion of a scene and focus for the thinnest possible lines with clear separation between them.

ION TRAP

In some very old TV sets there is an additional picture adjustment on the back of the TV tube neck—the ion trap magnet. Since this can seriously affect picture brightness, size and picture tube life, and furthermore, since its position near the picture tube socket makes it vulnerable to accidental dislocation, it is important to consider its purpose and indicate the method of adjusting it if and when required.

The ion trap magnet is a form of bar magnet held on the picture tube neck by a flexible coil spring. Its purpose (in nontechnical language) is to insure that the electron beam inside the neck of the tube is on course—moving properly along the axis of the tube. It is required because other electronic particles (ions) are incidentally generated inside the tube. Ions can damage a picture tube. The ion trap is thus part of a scheme of "separating the sheep from the goats"—getting rid of the ions, then making sure that the electrons move where they should be moving—through a number of small apertures toward the picture tube screen.

When adjusting the ion trap magnet a combined sliding/rotating type of motion is used. The object is to get a dual result—full picture screen coverage at maximum brightness. It is actually possible, through misadjustment, to illuminate only a central (circular) portion of the TV screen and at a low brightness level.

Since ion trap adjustments can be detrimental to the picture tube this procedure should be done carefully, repeated a few times, if necessary, until the best possible results are obtained. While a blank screen (no picture) is easiest to observe during this adjustment there is no particular objection to do this with any type of program on the screen.

In connection with the adjustment of the ion trap magnet, a note of caution is required. An improperly adjusted ion trap magnet, in addition to producing a poor picture, can also damage the physical interior structure of the picture tube, and will ultimately result in poor focus and dark spots (burn spots) at the center of the tube screen. To avoid this the magnet should be kept near the base end of the picture tube and adjustments made with the lowest possible brightness setting as quickly as possible.

For easiest adjustment and best results, it is recommended that the ion trap magnet be moved as far back (toward the tube base) along the neck as possible, consistent with obtaining the desired results.

The brightness and contrast controls must be kept low, so that slight changes are easiest to detect. After obtaining the best adjustment, it might be desirable to note or mark the position of the magnet both fore-and-aft as well as around the neck, with a marking pen or a small piece of masking tape; thus, should this adjustment accidentally be disturbed incident to other work in the back of the TV set, it is but necessary to move or turn the ion trap magnet to the marked location to restore the correct operation.

Ion traps haven't been used with new picture tubes in some years, but from time to time you may see an older set still operating with the original picture tube. New designs of picture tubes permits the deletions of the ion trap. If you own an old TV with a picture tube using an ion trap and purchase a replacement tube you'll find that you can throw away the ion trap. That is, new tubes don't need the ion trap.

MAGNET PRECAUTIONS

A warning pertaining to the effect of magnets, magnetic materials, and magnetic distortion on picture tubes is in order. While the subject of stray, unwanted magnetic effects on TV pictures came to the attention of the public only since the advent of color TV the problem has always existed, even though to a much less degree in all TV receivers. We shall return to this subject in connection with the matter of *color purity* in color TV sets. At this time we shall concern ourselves only with black-and-white TV sets and how they can be affected by stray magnetism due to the careless handling of magnets and magnetic materials in certain areas of the TV set.

As explained earlier there are a number of adjustments, such as width, linearity, etc., which depend either on the use of a magnetic material, or on the modification of an existing magnetic structure on or near the picture tube. Since iron and steel are magnetic materials (even if they are not magnetized, that is, they do not seem to attract nails, paper clips, etc.), they may, when brought near an existing circuit, change that circuit, at least temporarily. Since screwdrivers, pliers, wrenches, etc., are almost always of iron or steel their careless use on TV sets could adversely affect the TV set. For example, using a screwdriver blade to nudge or push an ion trap magnet or centering tabs is, therefore, not advised.

Similarly, using a pair of pliers to turn a tight thumbscrew, as used in some focus adjustments, can do more harm than good, particularly since the effect is not catastrophic but relatively minor. Of course, it is quite proper to use such tools on the chassis and other necessary areas in the cabinet, provided these areas are not too close to the tube neck, yoke assembly, focus magnet or the adjustments on the neck of the picture tube.

OTHER HORIZONTAL DEFECTS

Under the category of other horizontal defects come such abnormalities as:

- **Tearing**—Picture breakup into horizontal strips shifted to left or right, somewhat like Fig. 2-6 (Chapter 2).
- **Shifting**—What seems like a vertical split down the middle of the picture with the right-hand half on the left and vice versa, often accomplished by a very thin, faint picture (see Fig. 6-10).
- **Snaking**—The sides of the picture become wavy instead of straight up and down. This can also be accompanied at times by a rather thin, faint picture, although a normal picture may also be affected by this. See Fig. 6-11 for a typical example.

Fig. 6-10. Sketch illustrating horizontal shift.

■ **Streaking**—The picture seems normal in every way, except for dashed dark lines that seem to streak across the screen from left to right, somewhat like Fig. 6-12.

Referring to my description of the functions of a TV set I said that there are such sections as sync (short for synchronization) circuits, whose function is to amplify the precise timing signals from the transmitting station and use them for horizontal and vertical frequency control, thereby keeping the picture steady on the screen. A failure or defect in any of the sync circuits will cause a loss of control with some such consequences as listed above. To localize the defect make two or three preliminary determinations.

First, is the picture generally weak, faint, or thin, even if the sound is relatively strong? The fault may be in one of the amplifier tubes (other than sync). In other words any defective or old tube that causes a poor signal also causes a poor sync signal. While this condition affects both the horizontal and vertical picture stability, the vertical is far more susceptible.

Referring to Fig. 6-2 the offending tubes might be one or more of the following: V1, RF amplifier (on the tuner); V3, V4 or V5, IF amplifiers; and V6A, video amplifier. Replacing these tubes one at a time should locate the culprit. With the return of a normal contrast picture the sync problem will automatically disappear. Sometimes in the normal course of aging a *number of tubes* may be responsible,

Fig. 6-11. This picture distortion is called snaking.

each contributing partially to the trouble; thus, in a receiver with a set of tubes a few years old, more than one of the tubes listed may be responsible. Such tubes may test *weak-replace* or *?* on the tube tester scale. Replacing them one at a time will increase the performance of the signal stages, improving the sync and picture stability at the same time. If the picture contrast is normal the fault is most likely in the sync circuit, either tubes or adjustments or both.

Fig. 6-12. Picture streaking is caused by man-made noise, usually, from motors, appliances, etc.

Notice that, for example, in Fig. 6-3 there is a tube marked horizontal phase detector (V9A), in addition to the horizontal oscillator (V9B) and horizontal amplifier (V10). Similarly in Fig. 6-5 V7C is called horizontal AFC, while in Fig. 6-2 tube V10A is marked horizontal control. In each of these cases the tube referred to performs one type or another of horizontal sync function. A failure of any of the above mentioned tubes would invariably cause a loss of sync and some of the defects mentioned previously.

In those sets where a tube performs nothing more than a sync function the defective tube is easy to identify: removing it from the set (in parallel-wired sets only) does not affect the already malfunctioning sync circuit. In most sets, however, the sync tube or the control tube, as identified above, is one-half and sometimes even one-third of a multifunction tube (two-in-one or three-in-one tube). In these cases removal of the tube will also disable other functions, often leaving just a bright horizontal line at the center of the screen; however, this is of no consequence. Replacement with a good tube should restore the picture as well as normal functions. Incidentally, when testing a twin or triple tube in a tube tester the tube should always be discarded if one section tests below normal, even if the others are perfect.

Overloading

Paradoxical as this may seem too strong a signal can be just as detrimental to normal TV performance as too weak a signal. Specifically, overloading due to too strong a signal can cause snaking and picture tearing, and is usually caused by one or both of the following.

Contrast Control

Picture tearing can often result from excessive picture signal. Incredible as this may seem the technical structure of the average TV set is such that increasing the picture or contrast control to give a very hard or strong picture beyond that required for normal picture-viewing actually decreases the strength of the sync signal and often results in tearing. If this is the case, turning the same control toward a more moderate contrast setting should correct the tearing.

AGC Pots

In most TV sets manufactured in the past few years, there is a rear-chassis adjustment called AGC (automatic gain control). Its function, when properly set, is to place the receiver in such an operating condition that all stations in the receiving area will produce an acceptable picture with but an occasional adjustment of the front-panel contrast control; however, misadjustment of the AGC pot can cause no end of TV woes, not the least of which is loss of picture stability (hum, buzz, and even a complete loss of picture). Since most TV sets have a tube whose function is to regulate AGC performance a defect in the tube could also be responsible. Because of these possibilities the following procedure applies to the AGC stage and the AGC pot setting.

AGC Tube Change

With horizontal tear on the TV screen, remove the AGC tube and replace with a known good tube. Touch up the channel tuning. If the picture becomes normal momentarily, switch to an adjacent channel, then return to the test channel. If the AGC tube was at fault, switching channels should not matter. If tear still exists the AGC tube is not at fault.

AGC Adjustment

Assuming the AGC tube to be normal the adjustment is checked next. Using a moderately strong TV station, set the contrast (or picture) control about halfway, or a bit beyond (clockwise). Do not back off the contrast even if the picture seems too strong.

While watching the picture in a mirror advance (clockwise) the AGC adjustment on the back of the chassis until a buzz (or hum) is clearly heard. In some sets advancing the control too far beyond this point may cause a complete loss of picture. Exercise moderation.

Back off the AGC adjustment until all traces of the buzz disappear, or just slightly more than that. The AGC is now correctly set. Any malfunction of the horizontal sync will no longer be caused by the AGC setting, and other corrections must be attempted.

Horizontal Hold, Lock & Range

While all three of these adjustments may not be found in any one set, all have a horizontal hold adjustment, and some have one or the other of the remaining two. Any one of these may be responsible for picture tearing or instability.

Horizontal Hold—Assuming good tubes, a picture may fail to hold if the horizontal hold control is at either extreme of its range. It is permissible for the picture to begin to show tendencies of tearing or instability when the control is at either end of its travel. As mentioned earlier some horizontal hold controls have less than a full turn of adjustment range. In that case a fraction of a turn is all that will be required for correction. In slug-tuned horizontal hold controls, two or three turns in one direction or the other should correct any instability due to this setting.

Horizontal Range & Lock—These are always back-of-the-set adjustments and are intended to set the range over which the automatic horizontal synchronization is effective. The proper setting is that which will produce no loss in horizontal stability when adjusting the fine-tuning control (if any) or when switching from station to station. Starting with a condition of instability or tearing, the horizontal range or horizontal lock control is adjusted very gradually until the tearing disappears (picture is normal). Then a test for stability is made by switching channels or by turning the front-panel hold control to near extremes (almost fully clockwise, then almost fully counterclockwise). In either case there should not be a loss of sync.

Horizontal Shifting

This phenomenon is seldom present with a strong signal and it is likely to occur during freak reception; however, it can be caused by a substandard sync pulse, hence the procedure for sync tube changing or horizontal lock adjustment should be resorted to, if, upon careful thought, it is concluded that this is not a case of freak atmospheric conditions. Incidentally, a careful observation of the tube's behavior in a tester is of particular importance here. A *leaky* or *gassy* tube is quite likely to cause sync difficulties of every kind; so careful observation of the neon leakage test is imperative.

Horizontal Snaking

Tube leakage as indicated by a neon glow, and even a level of leakage not detectable by most commercial tube testers, may be responsible for this type of picture defect. Tube substitution, regardless of the tube tester's verdict, is the only way of discovering such a defective tube. In addition all tubes in the horizontal group may be responsible, even if they appear to be less likely suspects than the sync tubes. An incidental clue to such an elusive tube defect (not shown on the tube tester because the tube doesn't operate as it does normally) is the fact that the snaking appears only after the set has been operating for a relatively long time, when the tubes and their surroundings have reached their highest temperatures.

Streaking

This display is unmistakable and it is caused by an electrical noise (commonly called static on the ordinary radio) of the man-made variety, and there is no simple way to eliminate it. It may appear either as long dashed lines traveling slowly across the screen, or as a random band or stream of snow, also moving across the screen in unison. If it is chronic it may require relocation of the TV antenna by trial and error—a very costly remedy.

If the noise can be traced to its source (an arcing power company pole transformer, a heating system motor that sparks excessively, a similar workbench motor or some other defective electrical appliance), the proper procedure is to correct the condition at its point of origin. Should the source be a utility company device their cooperation will be required. It can usually be obtained provided a company representative is shown the evidence. There should be no difficulty in recognizing the malfunction and, if it is his company's responsibility, will arrange for correction. Where an in-house appliance is causing the noise, correction will have to be made by a competent servicer familiar with such appliances (oil burner, for example). More on this subject in the next chapter.

Intermittent Streaking & Picture Breakup

The tuner or front end can cause erratic, intermittent streaking, rolling, and tearing of the TV picture. This very critical and most sensitive of all sections of the TV receiver, black-and-white or color, is without question the most manipulated and most worn portion of the set. Regardless of make or model, the tuner contains a large number (up to 100 in some sets) of delicate switches or contacts, undergoing make-and-break operation whenever a station is changed. Since, as explained earlier the precise sync-timing signals are an integral part of the TV picture, the signal interruption caused by switching from channel to channel also removes the sync pulses. For a moment the TV receiver is on its own.

In a normally functioning TV tuner, that is, one that is not worn out or otherwise defective, switching channels momentarily interrupts the timing signals, but they are restored almost instantly. During the switching interval a quick tear or roll may take place, or there may be a momentary diagonal zig-zag. One frame (picture) may slide up or down. In a worn tuner, and that includes virtually every tuner after a few years use, the switch contacts tend to become intermittent and erratic, causing some misbehavior of the kind just described.

With a color receiver malfunctions such as color confetti, absence of one or more colors, and especially color instability and shifting can be due to poor contacts in the tuner. No special point is made of this fact under the color picture troubleshooting because the problem is not peculiar to color. In fact, it is much preferable (as is indicated in the introduction to color TV troubleshooting) to cut off the color when troubleshooting such color. When operation is restored to normal on monochrome color performance will automatically have been corrected. In other words this is a basic receiver problem, not a color problem.

There are two ways to solve a noisy or intermittent tuner problem: either thoroughly clean and lubricate it or have it overhauled by a specialist. A noisy tuner is so common in TV sets that a very practical and relatively inexpensive procedure has been established and is being followed by the repair industry. The old tuner is sent to a specialist overhaul shop where a factory-type overhaul is done on each tuner. The end results may be called as good as new. In view of the heavy use the tuner must endure, anything less than such a professional overhaul is a poor second choice, except where this service is not available or where a competent servicer is positive after examination that the overhaul is not required.

If you decide that cleaning and lubricating is the best course, you should begin, if you feel confident enough to tackle the job yourself. Start with a 100% detailed examination of the *before* conditions, including the method used to attach the tuner to the front panel or the chassis, and the exact positions, color, marking, etc., of each lead and cable from the tuner to the chassis. Most leads, in newer sets at least, are of the quick-disconnect type, but if one (especially wire braid or shielding) is soldered, carefully unsolder it, using no more heat than necessary. Of course, when you reassemble you must exercise equal care in resoldering to the same spot. Be sure the soldered joint isn't a cold-solder joint (a seemingly rigid mechanical joint with a grainy appearance that will pull off if tugged hard or pried). It is most important, also to return all wires to the same locations as before.

After removing the tuner from the chassis look at the contacts. Misalignment and looseness should be apparent now. In case of snap-in channel coils remove one coil to observe the interior. Using a fine, soft brush (a small camel's hair artist's brush is best) clean the wiping contacts and the shaft bearing with a contact cleaner sold in radio supply stores. Do not use household grease solvents, especially carbon tetrachloride. Allow all surfaces to dry. Best cleaning can be done while repeatedly switching channels back and forth.

TUNER SPRAY LUBRICANT

There is a quicker, although not necessarily better, way of correcting noise and poor contacts in the tuner without resorting to the more tedious removal, cleaning and lubricating procedure described previously. It is not a cure, but a correction which may last for quite a long time. It is practiced by many TV servicers whenever it is decided not to remove or replace the tuner. This method involves using spray lubrication.

Electronic component supply houses sell a pressurized spray cleaner/lubricant, marketed under various names, usually described as a cure-all or similar solution. The procedure still requires removal of the TV set from the cabinet and removal of the protective shield from the tuner proper. A small amount (so as no to cause dripping) is sprayed onto the contacts of the tuner switch sections while the station selector is manipulated, as when changing stations, to enable the fluid to reach all contact surfaces. The solution dries quickly and is claimed to leave no objectionable residue, only a film of conducting lubricant. Incidentally, removal of the selector knob and spraying into the tuner, as suggested by one spray manufacturer, is hardly a satisfactory solution.

Chapter 7
Internal Troubleshooting

In Chapter 6 I discussed what might be called external troubleshooting—the problems and solutions associated with the various controls and adjustments accessible from the outside of the TV chassis. In this and in the chapter to follow I shall concentrate on defects and cures originating beyond these adjustments, in the interior of the TV set. Since there is no absolutely clear dividing line between these origins and locations of defects, there will necessarily be an overlap between them; however, it is desirable to separate these, since the so-called external or adjustment type of defects are the most common and much more frequent.

BLOOMING & DEFOCUSING

This is a common phenomenon in older TV sets. A picture with blooming and defocusing problems exhibits insufficient brightness and contrast. Advancing either control, but particularly the brightness control, seems to have the effect of defocusing the picture, softening it as if a cloud were spread over the picture, fuzzing over the sharp details. The brightness may increase somewhat, but not enough before detail, and to a certain extent, shape is degraded. The fault lies in either or both of the following two areas.

High-Voltage Supply & Picture Tube

There is a considerable amount of interaction between the high-voltage power supply and the picture tube. In an old picture tube the electron emission capability may be substantially below normal. Increasing the brightness setting further taxes an already depleted source of electrons, while at the same time it upsets the voltage distribution in the high-voltage system, which often produces a decreased high-voltage (kilovolts) output. This is indirectly responsible for picture brightness. Similarly, the natural aging of some tubes in the horizontal deflection high-voltage system further acts to produce blooming and decrease electron beam acceleration.

Referring to the typical block diagrams in Figs. 6-2 through 6-5 in Chapter 6, we notice that each example has a high-voltage rectifier (V12 in Fig. 6-2, V11 in Fig. 6-3, V12 in Fig. 6-4 and V11 in Fig. 6-5), a damper (V13 in Fig. 6-2, V12 in Fig. 6-3, V13 in Fig. 6-4 and V10B in Fig. 6-5) and a horizontal amplifier (V11 in Fig. 6-2, V10 in Fig. 6-3, V11 in Fig. 6-4 and V10A in Fig. 6-5). In addition, in color TV sets there is also a high-voltage regulator which is covered in greater detail in a later chapter. Each of these tubes can be responsible for blooming in the following order of probability:

High-Voltage Rectifier

The high-voltage rectifier is almost certain to be one of the following tubes, regardless of the model number of the TV set: 1B3, 1K3, 2B3, 3AT2, 3A3, 1X2 or 2AS2. This tube is invariably inside the high-voltage cage or under a protective shield or barrier, as the case may be. As mentioned earlier it is difficult to see whether or not this tube filament is glowing. This does not matter in this case since the tube is certain to be on when the picture is blooming; the suspicion is that the tube is substandard. Replacement with a new one is a sure test. The ordinary tube tester is less reliable in this case since it may show the tube to be weak or good which is not good enough in this case. Standard precautions in handling this tube during replacement should be followed, of course.

Damper Tube

This tube may be indirectly responsible for blooming, but if the high voltage is substantially below normal, this tube may simultaneously affect vertical size, horizontal width, and the general performance of the set since this tube is an essential part of the overall power supply system. If the tube tests *fair*, or even *fair to good*, it is not good enough and should be discarded in favor of a new one.

Horizontal Output Tube

Since this tube and its circuits are responsible for producing the high-voltage supply for the picture tube, it can account for blooming because of reduction in this supply. While it was indicated earlier under decreased width in Chapter 6 that a poor horizontal amplifier tube will cause reduced picture width, this is not always the case. Depending on individual TV set circuit details, reduced high voltage *prevents* a reduction in picture width since blooming and dimming also expands the picture size, both horizontally and vertically.

Picture Tube

When no symptoms other than blooming and patchy, chalky white results from advancing the brightness and contrast controls, the picture tube is the most likely culprit. If the tube is a few years old this likelihood becomes a high probability. A picture tube rejuvenator or booster will extend the remaining life of the tube by a number of months. Care must be exercised to obtain the correct type of booster (for series or for parallel tube filament sets). Rejuvenation, a technique used by professional servicers, will accomplish the same purpose.

For three-gun color picture tubes boosters have become available and may be used, provided the instructions are carefully followed. Since the color tube consists of three separate electron-emitting structures, some boosters and rejuvenators have a simple provision for boosting one of the three, depending which of the color guns is deficient, by externally adjusting the booster before use.

High-Voltage Regulator

Because of the more critical requirements of the color system, additional circuits and tubes are used to keep the high voltage as nearly constant as possible. This applies to both color and black-and-white reception. Looking at the color TV block diagrams in Chapter 9 the regulators are V13 in Fig. 9-1 and V14 in Figs. 9-2 and 9-3. On color reception, blooming may evidence itself as color smearing, fuzzing and what is sometimes called color instability. In troubleshooting such a TV set, in addition to following the suggestion of turning down the chroma control to remove all color, a check after replacement of a suspected regulator tube should include turning up the chroma control to see how the remedial action has affected color reception.

DIM AND BLOOMING WHEN SET IS COLD

This behavior is exactly the same as described previously except that the picture gradually seems to achieve normal brightness and contrast after a good warmup period (a quarter-hour or more). The temporary condition is almost certain to be due to weak tubes. While a number of tubes can be responsible for this condition, the picture tube is the most likely suspect. The procedure, therefore, starts from this point; however, since it is impractical to remove the picture tube and test it like a small tube, a suitable booster is employed to obtain the same result.

Attach the booster to the picture tube, following instructions given in Chapter 8. Allow the tube to operate with the booster for a few days. If the trouble disappears, the picture tube is at fault. Three remedies are possible:

1. Leave the booster on permanently. It will prolong the life of the dying tube.
2. Have the tube rejuvenated. Approximately the same results will be obtained but the booster will not have to be used.
3. Have the picture tube replaced. Under normal conditions, between 80% and 90% of the tube's useful life has already been realized, so the cost of a new tube is quite justified if the TV set is otherwise performing satisfactorily.

If the booster does not materially improve the performance, a check of one or two other tubes in the set should be made. This includes the high-voltage rectifier and the damper tube, in that order.

If none of the above checks locate the cause there is the possibility that a component (most likely a filter capacitor) in the high voltage or B+ boost circuit is responsible. This, however, is not a task for a beginner.

HUM & BUZZ

These two names, hum and buzz, actually stand for two different symptoms; however, since it may sometimes be difficult for an inexperienced ear to clearly distinguish between the two, and further, since the two often appear together, I shall describe them together, but suggest how to identify their sources which may or may not be the same in all cases.

Hum

Hum can best be described as a smooth, steady low-pitched note that is the standard characteristic of household (60 hertz) AC. It can usually be heard on almost any tube-type radio with the volume control fully lowered if you place your ear close to the loudspeaker. It is almost always independent of the station being received, the weather, volume control setting, etc. In fact, it is the sound of 60-hertz AC when audible. In most tube receivers (and sometimes in transistorized sets when not operating on internal batteries) the hum is due to a small residue of this AC sound that is not filtered.

In severe cases the appearance of hum signifies one of three possibilities: First and most likely is that one or more of the filter capacitors in the receiver (radio or TV) has deteriorated with age and heat so that it does not function at maximum effectiveness. Since there is no simple visual indication or identification of the defective component, the services of an experienced technician are required. A second although less frequent cause of hum is a defective tube which has developed a *partial short* between the AC input and one of the functional elements in the tube. In technical parlance this is usually called heater-to-cathode leakage. This can be analyzed, located and remedied by following a simple test procedure.

Turn the volume up or down. If this seems to make no difference in the hum level, turn the volume fully off. If no change in the hum results (there probably will be no change) the fault most likely lies in one of the filter capacitors. Incidentally, this type of defect seldom happens suddenly, it is a gradual deterioration, not a catastrophic failure.

If the hum level changes with changes in volume a tube is quite likely to be the cause. Furthermore, such a defect is peculiar to tubes only; transistor circuits are not subject to this difficulty, although a filter capacitor failure as described could produce hum in a transistor receiver as well. To confirm your suspicion change to a different station and adjust the fine tuning. Any resultant changes in the hum level further point to a defective tube, and the isolation to a specific tube is carried out in one of two ways, depending on whether the TV set has a parallel or series heater string.

In the case of parallel wiring remove the sound discriminator tube. This is the first tube in the TV receiver which handles nothing but sound (audio). If the hum persists with this tube out of the socket, the fault lies beyond this point, closer to the speaker. This includes the first audio amplifier (if any) and the final audio amplifier.

To find out which of these two tubes is responsible replace the discriminator tube and remove the first audio tube. If the hum still persists the final audio amplifier is most likely responsible. Replace it with a new tube.

If removing the discriminator tube stops the hum, either that tube or one of the tubes preceding it is at fault. If replacing the discriminator tube does not cure the hum the fault lies in a preceding stage. In that case remove the first IF tube. The logical sequence here is the same as just outlined for the audio sections; thus, if the hum disappears with the first IF tube out of its socket, either that tube itself or one of the tubes from there to the antenna is responsible, including the RF amplifier (V1 in most sets) and the converter (V2). Remove the RF amplifier. If the hum stops this tube is at fault. If not, the next tube in the tuner (converter V2) is defective.

If pulling the first IF amplifier does not stop the hum, the fault lies in either of the remaining two IF amplifiers. Remove and replace the second IF amplifier. If the hum stops try a new tube. If not, the third IF amplifier tube is defective.

A final note on this hum problem. Earlier it was stated that while a constant hum regardless of the channel selector position or adjustment is most likely due to a defective filter component, and that a hum that responds to a change in station or volume control setting is probably due to a defective (although still functioning) tube, it must be pointed out that a tube-caused hum problem may produce effects other than audible hum. This includes snaking or weaving, or other image distortion on the screen; so be sure to observe any abnormal displays on the screen.

Removing tubes to trace a hum source is a convenient method for sets with parallel-wired filaments. But what about those sets with series filaments? Since removal of one tube (with rare exceptions) disables all others the only alternative is outright substitution all the way until the offender is found. Some time can nevertheless be saved here by proceeding as before from the audio output tubes back toward the antenna.

A general precaution in all tube-caused hum problems: after plugging in a new tube, if the hum seems to have gone, wait at least a few minutes before considering the problem solved. Many cases of tube hum (heater-to-cathode leakage) develop only after a thorough warm-up period and disappear after some cooling. Since removal and replacement of a tube lets all other tubes cool down, it is necessary to wait until all the (suspected) tubes again reach their operating temperature. Of course, this does not apply to parallel filaments, where all but the removed tube maintain their operating temperature.

Buzz

By this term I mean a sound characteristically peculiar to TV sets (except the very ancient ones having a split-sound system), and originating in the TV circuitry incident to normal operation but due to one or more misadjustments. This buzz is similar to the hum described earlier, and is of the same pitch (60 hertz) except for a raspy, buzzy quality instead of the relatively smooth tone of the hum. There is also another type of buzz, not related to the electrical operation of the TV set—a mechanical buzz. Although similar to the TV buzz, which is sometimes called intercarrier buzz, the mechanical type is more haphazard and random, and is easily identified. In either case buzz is a defect and requires correction.

Before attempting to correct this defect it is necessary to ascertain which type of buzz it is. The easiest to eliminate are the mechanical type buzzes, such as the buzzing from the loudspeaker or anything loose in or near the cabinet. A dangling wire, a tube shield not seated, or even a piece of paper or cardboard near the set can easily be identified, located and remedied by removal, tightening, etc. Loudspeaker buzz, although actually mechanical, is probably due to a defect in the speaker assembly—a tear in the speaker cone, a rubbing of an out-of-round voice coil (usually caused by warping due to heat) or even a simple loosened mounting. Any of these types can be recognized by comparison (keeping in step) with the audio, as well as total disappearance when the audio momentarily pauses. While a loose component is easily corrected, and a tear in the speaker cone can be judiciously repaired by application of quick-drying cement to the tear (not patching with tape), a rubbing voice coil is almost never satisfactorily repairable. But it is quite within the ability of the average do-it-yourselfer to replace the whole speaker by following a few simple directions.

1. Disconnect speaker from set. This will usually be a pair of wires connected by push-on clips, automotive style. Sometimes it may be a two-pin plug. Others will be soldered into the circuit. Dismount speaker from cabinet to gain access to cone (speaker front).
2. Carefully depress cone near its center so that it moves into assembly, at the same time listening and feeling for any

scraping or binding. There should be absolutely none. If this is the case the speaker is not at fault.
3. Obtain a duplicate speaker. Any parts supply house will identify the speaker and furnish a replacement. Of course, the physical dimensions (4 by 6, 5 by 7, 4 by 10, 6 by 9, etc.) should be the same, otherwise there will be a fitting problem in the cabinet. Make sure that the impedance rating is the same, that is, 4 ohms, 8 ohms, 16 ohms, etc.
4. Unsolder the old cable, if used, from the defective speaker and solder on to the lugs or terminals of the new speaker, taking care not to scorch or burn the cone in the vicinity.

If there is any question about the characteristics of the old speaker, a reference to the data package for the TV set (the parts list or the schematic diagram) will sufficiently identify the speaker to enable the replacement parts house to find a replacement.

The preceding examples were either mechanical types of buzz issuing from the offending part, or generated by the speaker in step with the sound. The following types are nonmechanical; but emanating from the speaker as does every other sound—music, speech or noise transmitted as part of the program. What is most characteristic about these types of buzz is their pitch; they are all of 60-hertz pitch, although their timber is different. It is not essential for the purpose of this explanation to consider the technical reasons for this; it suffices to say that both hum (usually called AC hum) and buzz peculiar to TV circuit operation discussed here are of the same basic frequency, which also is the frequency or repetition rate of the vertical sync. And although it was not intended that sync pulses reach the audio portion of the TV set, and ultimately the speaker, such is sometimes the case, as detailed below. When this occurs the audio system responds as to any other sound and reproduces it in audible form.

As to the difference in sound between these, the musician would perhaps explain this by calling the AC hum a pure (single) tone, and the buzz a complex tone, one having an admixture of many overtones. As in case of hum, there are some characteristics of buzz which help locate its origin within the TV set. These are discussed in the following paragraphs.

Warm-Up Buzz—During the first seconds (up to a minute or two) of the warm-up period, a buzz does not necessarily indicate any malfunction; stabilization of some of the automatic gain circuitry usually reduces this buzz to an inaudible level. No action is required in this case.

Constant Buzz—If the buzz remains for any length of time or is a constant annoyance simple adjustments may remove it. The checks for the specific cause should be made. Often it is necessary to set the fine-tuning control, not for the best sound consistent with a good picture, but for the optimum sound with good picture and minimum buzz. Once this control is so adjusted it should hold for most stations without individual adjustments. If this does not reduce the buzz to inaudibility, overloading may be responsible.

Assuming the tubes to be in good condition, that is to say, not very old, and the contrast control not set unreasonably high, check the AGC pot. This rear chassis adjustment regulates the overall amplification of the picture and sound IF stages. Too high a setting will produce a buzz and sometimes a glaring picture. As indicated earlier the AGC pot is adjusted by choosing a medium-good station. Adjust the contrast control about midway or until a good range of light and dark areas is obtained. Now advance the AGC pot until the buzz is audible; then reduce this setting just past the point where the buzz stops. Check the range of black-and-white areas to see that neither extreme has been lost, that is, the darker areas are not prematurely becoming a solid black while the very light areas are not becoming a glaring white. Adjustment of both the AGC and contrast may have to be made to obtain a happy medium—a full range of illumination from black through all shades of dim through full brightness with a minimum of audible buzz.

Sync Buzz—This fairly infrequent phenomenon of sync buzz is due to interaction between the sync tubes, vertical oscillator and amplifier and the video amplifier. Assuming that the TV set operated normally in the past, such a buzz may stem from two sources. One is the rearrangement (unintentional) of the wiring within the chassis so that the vertical sync and video amplifier circuits are too close to each other. This could happen only after a repair; therefore, it may be necessary to correct the *lead dress* to eliminate the buzz.

Practically every manufacturer has appropriate CAUTION notes to the servicer to replace any soldered lead in exactly the same position (as far as possible) as originally. We shall therefore assume for our discussion that any buzz pickup due to improper lead dress came about after a shop repair. The remedy is obvious. In case of the home do-it-yourselfer I can only reiterate that unsoldered-resoldered leads should be of the same length and in the same location as previously.

Another source of sync buzz is a malfunction in the video amplifier tube. Substituting a new video amplifier tube will prove or disprove this suspicion. Incidentally, a tube tester check of this tube may not be conclusive because a rather small leakage or amount of gas within the tube may be causing the trouble; and the defect won't show up until the tube warms up.

Intercarrier Buzz—This is a buzz inherent and characteristic of almost all TV sets except the ancient ones known as the split-sound types. In a great many TV sets, there is rear chassis adjustment (usually a slotted-shaft type) marked simply *buzz*. The control is rotated very gradually until the buzz is at a minimum. This should be done with a station tuned in while listening to the sound (music or speech) accompanying the program. In some TV sets there is also a tunable coil (called a quadrature coil) which is adjusted to eliminate buzz; however, this adjustment is seldom identified and it is best not to tamper with it unless you know what you're doing. Like any other tuned circuit adjustment, it cannot be properly made without professional equipment and skill. Fortunately, it seldom requires adjustment except after a repair in the immediate circuit vicinity.

MULTIPLE PICTURES

I refer here to the existence of two or more duplicate pictures one on top of the other. Usually each picture is not very stable and is only a fraction of the full screen height. The fault here is unmistakable: the vertical oscillator circuit is operating out of frequency. There are three possible causes, two reasonably likely, and the third very rare indeed.

Vertical Hold Misadjusted

Whether this control is an operating control on the front panel or a semiaccessible one behind a little door, it usually has sufficient adjustment latitude to cause oscillator operation at a fraction of its normal frequency or well above it. In other words it is possible to produce such a malfunction by rotating the control to its extremes. The first step to correct a case of multiple pictures is to adjust the vertical hold control first in one direction, then in the other. The picture will roll up or down, and after a while should jump from three to two and finally to one picture across the screen. Careful adjustment is required after a single picture is obtained to leave the control at the optimum position. This is checked by switching stations and observing whether or not the picture tends to roll.

Defective Component

A defective capacitor or resistor (changed value) in the vertical oscillator circuit will cause something other than one picture to appear on the screen. In rare cases, the vertical hold control can compensate for this change, but this is most unreliable. If the control has to be moved to an extreme position, a component is at fault and the *cure* at best will be very temporary.

Defective Tube

In very few cases a defective tube may cause this difficulty. If the picture can be stopped by setting the vertical hold control to its extreme position, you may try replacing the vertical oscillator tube; however, just a slight improvement, that is, requiring almost as extreme a position of the control for a single picture display, still points to a defective component.

Other Causes

What has been said about multiple pictures or scenes in the vertical direction, applies to some degree also to the horizontal direction; however, in case of horizontal off-frequency operation, either due to misadjustment or (more likely) to a defect, the whole image may be much less stable, going through any number of variations of Fig. 2-6, during which no recognizable picture may be displayed.

What is common to both types is the fact that there is a tendency for the scene to lock or stabilize at a sub-multiple of the correct repetition rate; thus, the vertical repetition rate is 60 hertz and the transmitted sync pulses normally insure locking at *exactly* this rate, provided the circuit is initially set as closely as practicable to 60 hertz. In other words, the transmitted sync information can and will correct minor deviations from exact frequency. Should this deviation be excessive, however, sync correction is no longer effective and the picture rolls or tears. Apparent locking or sync at a frequency which produces two or three pictures across the screen is what might be called a technical quirk (the explanation is beyond our scope here) occurring at exactly half or a third or sometimes even a fourth of the correct frequency. Should a malfunction cause the vertical sweep to operate at *approximately* 30 hertz, the sync will more or less hold, but two pictures will now be seen. Similarly, three images would appear if the frequency is around 20 hertz. Sync is less effective at each lower sub-multiple.

In case of the horizontal direction, the basic frequency is much higher (15,750 hertz). The number of the heavy stripes in Fig. 2-6 corresponds to the frequency the sweep circuit is operating.

The preceding explanation of multiple picture causes is intended to aid the do-it-yourselfer to analyze and locate the defect in such cases. Keeping in mind that by design the sweep frequencies are pretty close to the exact values, and further that the adjustments are but fine adjustments made to compensate for *minor* deviations due to normal wear and tear, it should be obvious to the troubleshooter that a severe or extreme condition cannot be *adjusted* back to normal.

Various suggestions given here, particularly the adjustments for purposes of correction, are intended and will be effective only in case of wear and tear deterioration or gross misadjustment through carelessness. In case of a component failure, whether a tube, transistor or passive part such as a resistor or capacitor, no amount of adjustment will effect a permanent cure.

Even in those cases where setting the control to or near one of its extreme positions of rotation brings back synchronization, the cure will only be temporary, since, as explained previously the setting of the control is only approximate, to enable the sync pulse to take over. With the control at its extreme, no correction latitude remains, except in the opposite direction of rotation. In other words, the control will no longer serve its normal function of being adjustable *up* or *down* from its center position. With but a slight further aging, no adjustment will be possible. Locating the defect and its correction is obviously the answer.

LINE NOISE & LINE NOISE FILTERS

In connection with the problem of electrical noise or interference such as streaks, dashes, etc., some words about noise filters are in order. There are many such devices or gadgets advertised to cure any and all TV ills. While they are not all completely useless, many of their claims are exaggerated. Before buying and installing one of these you should know what you can reasonably expect.

There are three sources of noise associated with a TV set, whether visual (in the picture) or audible. Two of these are external, while the third is internal. To dispose of the last one first, noise in a TV set can come from a defective component; an old resistor can sometimes develop what the TV servicer calls an intermittent, one with poor internal continuity. What is even more likely, due to the high operating temperature involved, is a tube may become a noise source. If it's a tube, it is often possible to locate such a culprit by gently tapping the tube with the rubber eraser of an ordinary pencil;

if the tube is noisy the tapping may produce flashing streaks in the picture or bell-like sounds from the speaker. Incidentally, a tube tester is of little value in performing reliable tests for noise. In case of a defective component, a professional will have to be called upon for analysis and repair. The same holds true for operating controls which become noisy with age. This includes contrast, volume, tone, fine tuning and channel selector.

Line noise includes all types of electrical noise generated by appliance which operate on AC. While not all appliances are potential noise makers, some of them are. Their electrical noise travels along the AC wiring in the house and enters the TV set. By contrast noise from utility poles equipment and devices outside the house seldom reach the TV set this way. But this is nothing new, and radio and TV manufacturers have, almost without exception, provided simple noise filters inside the receiver for this purpose. Nevertheless, it is possible that an additional filter may be of some help if it is the proper kind and is properly installed.

It is very doubtful whether the simple little gadget that looks like an AC plug on one end and an AC receptacle on the other can be much use. The fact that the built-in filter does not seem to help suggests one of two possibilities. Either the noise does not enter via AC line, or if it does, that the elimination of the noise must be accomplished at the source, that is, at the offending appliance, be it the washing machine, oil burner, electric drill, etc.

The third source of noise, and a very likely one, is the antenna. Of course, an old corroded antenna with poor lead-in connections is a potential source of noise. But what I have in mind here is that noise generated elsewhere and picked up by the antenna along with station signals. Other than relocating the antenna and transmission line, nothing can be done here, except enlisting the aid of the utility company if their power lines, pole transformer, etc., prove to be the cause. In fact, for any but the simplest type of local noise the average TV owner is hardly in a position to move his or her antenna any appreciable distance, especially horizontally. A change in the antenna height may be helpful if it can be first found experimentally that a different height (higher or lower) proves noise free.

With regard to the transmission line the problem is easier to solve. If it is apparent that the transmission line runs along a noisy structure or device or power line, either relocation or shielding should produce some improvement. Relocation is simpler and should be tried first. If this is not feasible or it does not help, substitution of a shielded (metal clad) transmission line is almost 100% certain to be effective.

There are two simple types of shielded transmission lines suitable for practically all TV sets. One is unbalanced 75-ohm line, sometimes called a coaxial or coax line. This contains an insulated center conductor and a metallic (braided) outer conductor, sometimes rubber covered. When installing this lead-in the outer conductors must be thoroughly grounded by a solid connection to the TV chassis. In addition the TV set must be equipped (some are) for a 75-ohm transmission line. Otherwise, a little accessory known as a matching transformer must be connected between the coaxial line and the TV set.

A second type of shielded transmission line is the 300-ohm twin-lead shielded line. This is the usual two-wire flat line with the addition of an external shield. Assuming that the TV set requires a 300-ohm antenna (most sets do), the connections are the same as with the unshielded transmission line, except that the external shield must be tightly connected to the TV chassis at one end and to the TV mast at the other. There's usually a slight signal loss because of this outer braid or shielding, but this is not significant, provided the antenna proper is fairly good.

A final note on noise prevention or elimination. The notion that an indoor antenna is immune to outdoor noise is completely false. No compact, simple, abbreviated antenna can improve reception over a good outdoor antenna. In fact, the *window improvements* can seldom approach the performance of even the simplest outdoor antenna properly installed. Except for locations with extremely high signal strength, these window gadgets are fairly useless. And as to devices which employ radar principles and convert one's housewiring into one giant all-direction antenna, the best that can be said for them is that they will work where any scrap of wire will work; they will not be much use where a normal antenna is required.

Chapter 8
Troubleshooting Catastrophic Failures

Up to now the various adjustments and corrections I've covered have been primarily of the preventive-corrective maintenance type, intended to keep a TV set in normal operating condition, as well as to correct deviations from such normal operation due to aging, wear, deterioration, etc. Our concern now involves catastrophic failure where the TV set ceases to operate either partially or totally.

COMPLETELY DEAD RECEIVER

As obvious as this may seem, it is sometimes overlooked. If there is absolutely no evidence that the power is on (pilot light on, some tubes lighting in transformer type sets, etc.), a few simple tests will provide the answer. They should be made before going further.

- Check for a loose plug, or a disconnected plug at the wall outlet.
- Check for a defective wall outlet by plugging in a lamp or other appliance.
- Check for a defective fuse or open circuit breaker. In the earlier sets a power line fuse can be found on the rear apron of the TV chassis (the metal strip at the bottom of the set). More recent receivers are likely to have a resettable circuit breaker that may be reset or closed by a simple push. This requires a few seconds for cooling off after the

set goes off. Should the circuit breaker pop open again immediately after resetting (or, for that matter, should the replacement fuse blow similarly), there is a defect, such as a severe overload or short circuit, and further troubleshooting is required.

■ Check the on-off switch on the TV set. It may be defective.

If the answer to the first two checks is negative for transformer type sets, the fault lies either in a defective on-off switch or a defective power transformer. In either case the repair is not for a beginner. The switch is almost always part of a control assembly (volume, tone, etc.) and requires professional attention. A transformer replacement is even more out of the reach of an amateur. Do not attempt either. Fortunately, both of these possibilities are rather improbable.

In a transformerless (series tube hookup) TV set the next step is to proceed according to the detailed instructions given in Chapter 5. Since any one tube (or tube group, see Figs. 5-3 and 5-4) or fusible resistor will stop all tubes from lighting, the procedure suggested will almost certainly lead to the open circuit. The remedy is then obvious.

TUBES LIGHT ONLY

There are two inferences to be drawn from this symptom. One, the fault lies in a part of the set common to **both** video and audio. Second, since the screen is dark, a fault common to both the signal circuits and deflection circuits is responsible.

A power supply failure may be either very simple or beyond the capability of the beginner, depending on whether the power supply uses tube rectifiers (V14 in Fig. 6-2, V14 in Fig. 6-3) or semiconductors (selenium or silicon diodes as in Figs. 6-4 and 6-5). In the case of the latter (and these also fail most infrequently, usually due to gradual deterioration with age), a professional's skill is required, even if it is a quick and simple job.

RASTER ONLY

This symptom eliminates the horizontal and vertical deflection circuits and points strictly to sections common to the video and audio portion of the set. Of course, this failure rarely exists in transformerless sets, as any tube failure usually disables everything—video, audio and raster. Referring to the block diagram in Fig. 6-3, all tubes in the tuner (V1 and V2), IF amplifiers (V3A, V3B and V4) and the video amplifier (V5A) are suspect. Sound IF tubes (V5C, V7A and V7B) are not likely to be involved, even if the sound is missing because the simultaneous absence of the picture points to one or more of the tubes from the video amplifier (inclusive) back toward the antenna. By substituting tubes as suggested earlier you should quickly locate the problem if it is a tube.

NO SOUND BUT PICTURE OK

Failure of the sound may involve any of the stages and tubes beginning with the sound IF, just following the video detector, up to and including the speaker. This includes V5C, V7A and V7B in Fig. 6-3, V5A, V6 and V7 in Fig. 6-4, etc., and applies equally to parallel and series filament tube sets, since the presence of a picture precludes any tube burnouts. Sound failure can also be due to a failure in the sound detector, which in some sets consists of a pair of semiconductor diodes; however, diode failure is most infrequent and may be anticipated as a wear-out type of failure, taking much longer than tubes. In those rare cases when the sound is weak even if all the tubes identified above test good, the detector diodes may as a last resort be suspected.

There is an important caution that must be made in connection with the sound section of a TV receiver and it pertains to the handling of the speaker. Since most speakers on TV receivers have quick-disconnect clips or pin-and-jack connections (for convenience in removing the chassis from the cabinet) it is quite possible to deliberately or accidentally pull the speaker leads off, leaving the speaker electrically disconnected from the receiver. This is potentially damaging.

In tube-type sets disconnecting the speaker with the set on may damage the output tube. With transistor sound sections, disconnecting the speaker with the set on spells sure *death* to one or two expensive transistors; therefore, *NEVER* disconnect the speaker leads while the TV is on. In the troubleshooting procedure just discussed, first switch off the set, substitute for the suspected tube, then switch the receiver on again. While damage is not 100% certain in some tube sets, it is not worth risking such damage. With transistors damage is certain.

Now and then there will be a TV set (or for that matter any solid-state radio or record player) in which the power transistors are said to be protected from such damage. My advice is still valid—play it safe and do not disconnect the speaker while the piece of equipment is on.

Although extremely rare now, in some TV sets, known as the split-sound type (as distinguished from the current intercarrier type), there may be a sound failure in stages not identified on the block diagrams in Figs. 6-2 through 6-5. As an illustration, Fig. 8-1 is a block diagram of an old split-sound TV set. Notice that the sound and picture information channels split apart (hence the name) immediately after the tuner. In such a set, if any are still around, the absence of sound must be due to a fault in V9, V10, V11, V12, V13, or V14. None of the tubes in the upper string (the video IF), including V3, V4, V5, V6, V7 or V8 can be involved. Incidentally, the two (or three) tubes in the tuner, V1 and V2 (and V3), cannot be responsible for the absence of sound unless both sound and picture are missing.

In connection with this type of relic it should be pointed out that a failure of picture only, but with a raster present, must be charged to the upper leg of the split, that is, the video IF channel comprising V3, V4, V5, the video detector, V6, and the video amplifiers V7 and V8.

SNOW

This picture defect, snow, may vary from good sound, snowy picture to fairly good sound, very faint picture, or no picture at all. In addition, the snow may appear at random, suggesting (improperly so) that the fault lies outside the set. Actually, this is not so. Snow is a name given to a picture which seems to be broken up into small flecks or pieces which seem to shimmer or waver. They do not fall or move down the screen, although in a general way they are reminiscent of snow-filled air.

Regardless of which of these snow problems a particular TV set has, it is distinctly different (in fact, opposite) from the weak picture cases described earlier. In the vast majority of cases the presence of snow is evidence that the TV set is working hard to produce a picture from a poor signal. This means that either no signal is received by the TV set, or the received signal cannot get through due to a break in the functional continuity in the set. The procedure for isolating the fault and applying the appropriate remedy is simple.

Examine the antenna lead-in system. An old or otherwise poor antenna system will not necessarily cause a sudden appearance of snow. But a break in one or both conductors of the lead-in at the point of connection to the antenna, at the point or points of fastening on the way down, at the entrance through a window or wall or finally

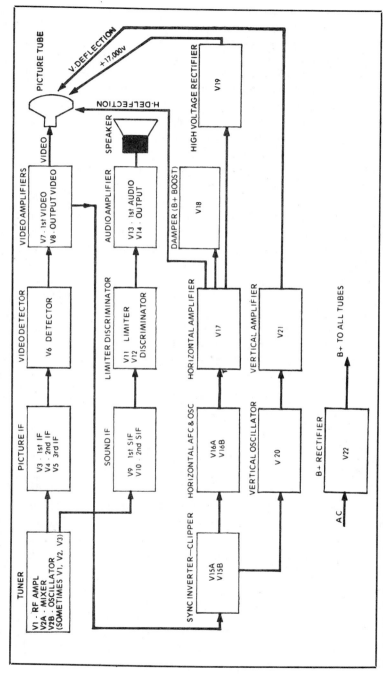

Fig. 8-1. Block diagram of an older split-sound TV receiver. Notice the separate IF stages of sound and picture, hence the name.

at the connection to the antenna terminals on the rear of the set—a break at any one of these vulnerable points is a likely cause for snow on an otherwise normal TV set. To check for a suspected broken lead-in, a substitute antenna, rabbit ears, an indoor wire antenna, or even a length of wire connected in place of the suspected antenna will be sufficient to verify whether or not the antenna is at fault. Of course, allowance must be made for the fact that the substitute antenna is, at best, a mediocre performer. But if the picture improves and the snow effect is diminished, the antenna lead-in is probably at fault, and that means a further check for a break in the wires. If, however, no appreciable improvement is noticed with the substitute, the antenna system is not at fault.

Check the RF amplifier tube (by tester or substitution) in the tuner (V1 in all block diagrams). This tube, more than any other, is likely to be the culprit. Even when it has deteriorated very severely this tube will still allow some signals to get through; hence, it may not be suspected. But the selfcompensating functions of a modern TV set will try to make up for the severe reduction in signal, amplifying a lot of noise which appears as snow. If no improvement is attained, similarly check the first IF tube (V3 in all block diagrams). To a lesser degree than V1, this tube is also potentially responsible for snow and should be checked accordingly. Check the remaining IF tubes if the RF and first IF tubes seem *OK*. While the probabilities are much lower that one of these tubes is reponsible, they should be checked in a tester or by substitution.

While the tuner is a most likely suspect for this defect it is assumed that tuner is operating normally since, as discussed earlier, tuner abnormalities evidence themselves in other ways and their repairs will automatically preclude them as a source of snow. Nonetheless, and especially in cases of minor deterioration in tuner operation, which many TV owners are willing to tolerate for a while, imperfect contact in one or more of the tuner switch contacts will also cause a snow effect on the screen. Usually, in the early stages of tuner wear, it is but necessary to jiggle the tuner channel selector to obtain a better contact and eliminate the snow. Incidentally, in such a case this is an obvious hint that the tuner needs maintenance—cleaning at least.

SOUND OK BUT PICTURE MISSING

A totally blank screen with normal sound points to trouble in the video portion of the TV set, including the picture information and horizontal sweep circuits (if there's no raster) but not the vertical or sync circuits. The horizontal stages are involved because they are responsible for the presence or absence of any light (raster or picture) on the screen.

The troubleshooting procedure is based on some logical assumptions. First, the picture tube is not burned out. In a transformer-type set, a burned-out picture tube is identified like any other burned out tube—absence of any light near the base of the tube. Of course, it is assumed that the picture tube socket is properly seated on the tube base. Once in a long while it may work loose or simply lose proper contact. Second, the tube did not become defective since the last time it was on; it usually doesn't happen that way, certainly not suddenly. If the tubes are wired in series and all tubes light, the procedure is no different than for a parallel tube set. If no tubes light (remember, this is an all-or-none situation) the procedure outlined previously should be followed until the bad tube is found. If this does not solve the problem, that is, tubes will light but still no picture, the troubleshooting is the same as for the parallel filament sets.

The step-by-step sequence for locating the cause of a dark screen is easily followed. Replace the horizontal output tube (V11 in Fig. 6-2, V10 in Fig. 6-3, V11 in Fig. 6-4, V10A in Fig. 6-5), etc. Since this tube is directly responsible for generating the high voltage (15,000 to 25,000 volts) necessary for producing light on the screen, failure of this tube will result in a totally dark screen. Precautions outlined previously regarding the handling of this tube, especially if it happens to be inside the high-voltage cage, must be followed at all times. If a new horizontal output tube does not solve the problems, the next step is in order. This is also assuming that the horizontal output stage is not working. although the tube itself is okay.

Carefully remove and replace the damper tube (V13 in Fig. 6-2, V12 in Fig. 6-3, V13 in Fig. 6-4, etc.). This tube, if faulty, breaks the path of the B+boost (250 to 650 volts), again disabling the horizontal circuit responsible for producing light on the screen. In most cases this would have to be a catastrophic failure; a weak or even poor testing damper tube will not remove all light from the screen.

If the damper tube proves to be good, proceed to carefully remove the high-voltage (15,000 to 25,000 volts) rectifier (V12 in Fig. 6-2, V11 in Fig. 6-3, or V12 in Fig. 6-4). This tube is always inside the protective cage and it always has a top cap connection which may be reluctant to come off. An additional dificulty, even if a minor one, is the fact that this tube lights very dimly, sometimes almost invisibly (depending on its physical position) and hardly feels warm to the touch, hence should not be assumed to be bad just by appearance alone. Do not try to feel this tube while the set is powered.

Incidentally, this is the only tube in a transformerless (series filament tubes) TV set that is not part of the series string; it can be burned out even if all the other tubes in the set are lighting. The foregoing procedure, one or more tubes should restore the picture or at least the raster to the screen; however there is one other possibility in some makes of TV sets.

As stated earlier some TV sets make use of an internal fuse, usually inside the high-voltage cage, for the protection of the horizontal output portion of the TV set only. This is not the main fuse which protects the whole TV set. When this internal fuse blows, the whole horizontal output circuit, including the high-voltage system, is interrupted, resulting in no light on the screen. Replacement of the fuse is relatively simple, even in those cases when the fuse is soldered into position by pigtails.

Visual inspection must be very carefully made as the metal conductor inside the glass fuse is very thin; therefore, an open is not too obvious. In replacing the fuse, if it is of the snap-in type the procedure is quite obvious. The fuse should be of the same physical length in order to fit into the fuse holder clips. Its electrical value is very critical. If the fuse is of the soldered-in type a slightly different replacement procedure is called for. All radio repair supply houses sell a replacement fuse assembly which looks somewhat like Fig. 8-2B; as can be seen, this is a double fuseholder or two fuseholders fastened back-to-back. One side is slipped over the old presumably burned out fuse, while a new snap-in fuse is placed on the opposite side.

DIM PICTURE NOT COVERING FULL SCREEN

When operation is just a little below normal, that is, when the picture could be a little brighter and have a little more contrast and when the picture does not quite reach the edges, the fault is probably a poor rectifier tube (or tubes). Referring to the block diagrams again (Figs. 6-2 through 6-5) we find the rectifier tubes identified as 5U4G, 5U4GB, 6AX4, 5AX4.

In some very large TV sets, particularly of earlier vintage, two rectifiers (a pair of the same or two different types) are used. Since in every case the rectifier supplies the power to all other tubes in the

Fig. 8-2. Some fuses are intended to be soldered in place; therefore, they have pigtail leads (A). Solderless fuseholders (B) are available. They clip onto the burned-out fuse and the replacement fuse is inserted in the clips on top of the fuseholder.

set, a reduction in output here will reduce the performance of most tubes in the set to some degree. Furthermore, rectifier tubes operate at high temperatures and full load at all times and are more prone to deteriorate than any other tube in the set. Replacement of weak rectifier tubes usually corrects insufficient brightness, poor-contrast, and insufficient height and width.

It should be mentioned here that there is an auxiliary rectifier tube, previously identified as a damper or B+boost. While the damper does not affect all the tubes in the TV set, it directly affects the horizontal and vertical circuits, hence the height, width, raster and, indirectly, picture brightness.

PICTURE DIM & BRIGHTNESS/CONTRAST INEFFECTIVE

There is another case of improper brightness or contrast control where the picture size is normal. Unlike the previous example where the brightness and contrast controls are functioning but at their maximum settings there is still not sufficient picture brightness, this case identifies a condition in which the brightness and contrast controls are not functioning properly. Sometimes increasing these settings, or advancing the controls, will produce a photographic negative effect—blacks are white and vice versa. More often the gradual variation from dark to light to bright, as is the case with the average TV scene, will not be obtainable, except perhaps at very low (dim) settings of the brightness and contrast controls. At all other positions, as when attempting to get a normal picture, the light and dark areas appear flat, muddy and very dull. Stmetimes such a picture suggests looking through a very dirty window or through a gray filter at a normal scene. Any of these symptoms, alone or in combination, suggest a defective picture tube (of course, it is assumed that the problem is not the one just described previously). The electron emission has deteriorated to such a low fraction of the normal amount that internal adjustments or small tube replacements will no longer help. The only permanent remedy is the replacement of the picture tube, and this is a job for the professional servicer.

If you're shopping for a replacement picture tube (plus installation) some money can be saved by buying a rebuilt tube. In fact, if no specific instructions are given to the contrary, the replacement tube may well be a rebuilt one. This is quite satisfactory, both ethically and technically, provided it is a tube rebuilt by one of the standard tube manufacturers. The 100% brand new tube today is still available, but its use is the exception rather than the rule, and it offers absolutely no advantage over the standard rebuilt tube.

When it has been established that a dim picture is due to a weak picture tube, the life of such a tube can be extended, often for a number of months, by a simple technique commonly called rejuvenation. What it amounts to is a rejuvenation of the electron-producing element of the tube. It can be restored, for a while at least, by the application of a higher-than-normal voltage, raising the temperature of the electron-emitting surface on the tube element.

Most professional servicers use a one-shot remedy by applying an overvoltage for a short while, thus reactivating the electron emitter. After this, the tube reverts to its normal operating voltage. You may also install a booster device which is attached to the tube and left there for the remaining life of the tube. It is simple, less expensive, effective, and most important it may be less likely to shorten the tube life. Since a rejuvenation overvoltage is applied for only a relatively short time, it is necessarily more drastic and may have some delayed action effect on the remaining life of the tube. By contrast, the booster, which continuously operates with' the tube, applies a much more modest overvoltage and is, therefore, less detrimental to the remaining tube life.

Figure 8-3 is a sketch of a common picture tube booster. At one end is a socket (or section of one, as found on most TV sets) which fits over the picture tube base, exactly like the original one. At the

Fig. 8-3. Sketch of a typical picture tube booster.

other end is a plug which contains a small transformer. Connection to the TV set is simple.

Switch off the set and carefully pull the socket from the picture tube base. Mate and connect this socket with the base plug on the booster. Connect the socket on the other end of the booster cable to the base of the picture tube. The booster may be left hanging as is.

One important reminder: there are two different types of boosters for sale in radio supply stores and they are not interchangeable, even if they look alike. One is exclusively for parallel (transformer-type) tube hookups and the other is exclusively for series tube hookups (transformerless). Make absolutely sure which type of TV set yours is and purchase the correct type. Some types are for both filament arrangements by virtue of a moveable wire connection.

HEAVY BLACK-AND-WHITE BARS

This symptom, heavy bars, looks somewhat like Fig. 8-4A. There may be either one dark and one light horizontal bar, each covering approximately 50% of the screen height, or, as in Fig. 8-4B there may be three bars, one wide and two narrow. Often these bars will slowly drift up or down the screen. To locate the cause of this malfunction, more than one step is usually required.

Understanding the problem will make the solution so much easier. The vast majority of such bars is caused by what is commonly called AC hum, although the word hum usually refers to the audible manifestation of the unwanted presence of AC in a circuit. One of the most likely sources of such AC hum (or at least those within the ability of the beginner) is a tube.

A defective tube in this case is one which has developed what is called heater-to-cathode leakage. The technical significance of this statement is of no particular importance at this point, except insofar as corrective action is concerned. Specifically, when one or more tubes are suspected of causing the horizontal bars, you should test them very carefully on a tube tester having a leakage indicator, usually in the form of a small neon bulb. Total emission tests, as indicated by the needle swinging into the replace, weak or good portion of the scale, are of no value here, and may, in fact, be misleading, since tubes with heater-to-cathode shorts or near shorts often read high.

Not only must the neon glow be looked for very carefully, but enough time (at least a few minutes) must be allowed for the suspected tube to reach a high enough temperature to produce the leakage or short. The same precaution applies to the tube substitution procedure. Merely switching off the TV set often allows sufficient time for the culprit tube to cool somewhat and operate normally when the set is switched on again.

Therefore it is imperative for the beginner to follow two simple steps in order to eliminate this ambiguity of indications. First, observe how soon the bars appear after the set is turned on initially. Second, when switching tubes allow at least that much time to see whether or not the bars reappear before concluding that the job is done. It should be noted here, that contrary to most other malfunctions in a TV set, which usually can be traced to one particular and exclusive tube or stage, the AC hum responsible for the bars under discussion here can originate almost anywhere in the TV set, although under different circumstances and with different symptoms. Because of this, the troubleshooting procedure must be able to locate the defect without a random hit-or-miss, trial-and-error, let's-see-what-happens type of procedure.

To locate a defective tube, it is most helpful to isolate the fault to a particular portion of the TV set, keeping in mind the fact that a number of tubes can be removed from the transformer-powered TV set without affecting the raster. For example, referring to a typical block diagram as in Fig. 6-3, all the tubes, individually or in a group from V1 through V7, can be removed without removing the raster, although the video picture itself may disappear. A systematic isolation procedure for this TV set is very simple.

Remove the first IF (V3 in most sets); both picture and sound will be disabled but not the raster. If the bars disappear, the defect is

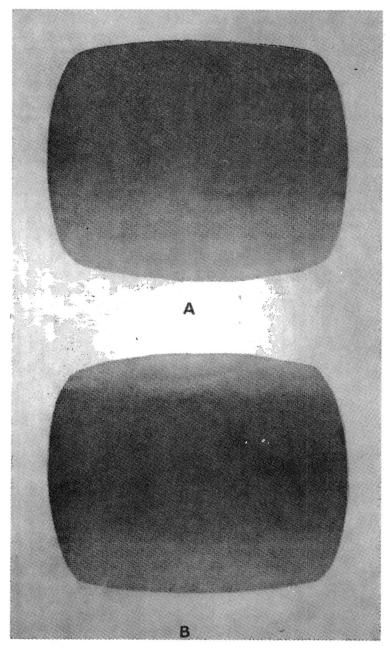

Fig. 8-4. AC hum can cause the raster to vary from black to white (A) or from white to black to white (B).

either in the tube just removed or in the tuner itself. To determine which is at fault, replace V3 and remove V1. If this removes the defect, V1 is at fault. If the bars persist, the fault lies either in the converter tube, V2, or in the previously removed V3. If neither of these tubes affect the bars, the trouble lies further on, in the IF stages, V4 and V5, the video stage, V6, or in the picture tube itself, as we shall determine presently.

Remove the last IF tube (V5 in Fig. 6-2). This also disables the picture and sound, but since the video amplifier (V6) is still in its socket, it is a most likely suspect if the bars remain. If the bars disappear, the fault is somewhere in the IF circuits, such as V4 or V5. To determine which, replace V5 and remove V3. If the bars disappear, V3 is at fault; if the bars persist, V4 is at fault. If removing V5 makes no change the IF is not at fault. If the bars persist the video amplifier is almost certain to be responsible.

Another possibility, although a much less likely one, is the sync tube(s) should be checked. In fact, after eliminating the video amplifier as the cause, the sync tubes become the prime suspect.

Remove V6A, the video amplifier. This will disable both the picture and sound, as was discussed earlier. If removing the video amplifier also removes the bars the logical step is the replacement of this tube. If this cures the malady this tube was at fault. If the bars persist remove and replace the sync tube, V8A. If the bars are still present the fault lies beyond this point either in the picture tube or in the power supply to the picture tube.

Very often the fault will be in the power supply portion of the picture tube circuits (that is, filter capacitor) but most likely the picture tube itself has an internal short or a near short. Occasionally, even if rarely, it is possible to burn out the short with a tube rejuvenator. If a professional servicer offers to do this, he or she will probably stipulate that it must be done at your risk. This is not unreasonable and entails no loss to the owner, since the tube is not serviceable and its trade-in value will not diminish even if it is burned out in the attempt.

PICTURE SNAKING

This defect in the TV picture is mentioned here because it is usually caused by the same type of tube defect as the AC bars just discussed, namely, leakage between elements in one of the tubes. All vertical lines—objects, picture edges, or people—seem to weave in shapes of S-curves. Almost any tube in the TV set may be at fault, but it's most likely to be tubes in the horizontal, sync and video circuits (V10A, V10B, V11, V8A and V6A in Fig. 6-2), and least likely to be those in the tuner (V1 and V2) or in the IF amplifiers (V3, V4 and V5). The procedure for locating the bad tube is the same as for the just discussed hum bars.

SOUND BARS

In connection with this phenomenon, mention should be made again of this type of defect stemming from two other sources (overloading, AGC adjustments, etc.). The first is internal and is under the control of the user, the second is atmospheric, and virtually nothing can be done about it. Advancing either the contrast or the AGC adjustment beyond their normal settings as described in Chapter 6 will give rise to various degrees of instability and even snaking. The remedy is obvious.

In case of the external cause, little can be done. Normal TV reception is based on line of sight signal travel from transmitter to receiver; that is, the FCC allocated station locations and power outputs is such that two stations of the same channel number are separated geographically so as not to interfere with each other. At infrequent times, however, and often incident to some radical weather change, signals travel not only directly (line of sight), but also via reflection from an electrically charged layer above the atmosphere, covering much greater distances than intended. Such a signal will play havoc with the local station of the same channel (since there are minute differences between the two stations despite the fact that the same FCC standards apply to both), affecting sync, AGC and other vital functions. The effect may differ from set to set and from station to station (the higher frequency channels are usually not affected), but while it lasts little can or should be done about it. Any attempt at correction through some adjustment will simply misadjust the set so that when conditions return to normal the set is actually out of correct adjustment.

One final possibility need be mentioned—in overloading due to the proximity of a powerful local station. Although relatively rare, this has happened in the past. In fact, an offending radio station actually distributed little devices—filters, of a sort—to residents in the vicinity of the powerful transmitter, in order to eliminate or at least minimize the problem. There are simple little pads for sale, actually signal reducers, which connect to the antenna terminals of the TV set. Of course, the simplest remedy, even if not 100%

effective, is to reduce the contrast control when the particular station is used.

Sound bars are somewhat similar to the heavy black-and-white hum bars except they are much thinner (Fig. 8-5), more numerous (there may be a dozen or more), they appear somewhat wavy, and, instead of slowly drifting up or down, they seem to waver with the sound form the speaker, as if animated. Sound bars appear when some of the sound energy reaches the picture tube and is reproduced as light. The cause of this malfunction is seldom a defective tube, rather it is due to an improper adjustment either of the fine-tuning control or, more seriously, one of the tuned circuits inside the TV set; however, only the fine tuning is within the capability of the beginner to correct. Try to adjust the fine-tuning control for a clean picture even if it means some reduction in volume.

If this is not effective, it is an indication that either one or more of the IF circuits is improperly adjusted, or, in some TV sets, the sound trap (called 4.5- MHz trap) circuit is improperly adjusted. This should not occur if the TV set ever functioned properly, as it is extremely unlikely although possible that these circuits would drift out of adjustment; however, after a repair, or especially alignment in the repair shop, some of these circuits may have been improperly adjusted.

HERRINGBONE WEAVE

A herringbone pattern on the screen is seldom due to internal causes, but these cannot be automatically discounted. One likely cause of this phenomenon is interference from another TV set. Although the FCC has established some very clear standards for TV receiver radiation, many sets will make their presence known in other neighboring sets, especially in apartment houses and in master TV antenna installations. This can easily be identified by its random, intermittent nature and absence from most channels. It appears only when the offending receiver and the victim sets are tuned to the same station. There is little that can be done about it, other than attempt to minimize it by touching up the fine-tuning control or switching to a different channel for the duration.

Where there is an adjustable fine-tuning control, a herringbone pattern may be due to improper setting of this control. In older sets, adjustment of this control is required each time the channel is changed. The proper setting is the point which gives the best picture, not the loudest sound (these two were not always coinciden-

Fig. 8-5. The effect of sound bars is similar to hum bars, except there are many more of them.

tal). In more modern receivers where the fine tuning is not critical and where usually one setting serves for all channels, there is considerable leeway in setting this control, but it can still be set (improperly) so as to produce a herringbone weave on one or more stations. An optimum setting (good for all stations, not best for any one) should be made by trial and error to avoid herringbone effects.

Incidentally, since many TV sets may receive the same channel on two adjacent selector positions (that is channel 2 on dial positions 2 and 3, channel 7 on positions 7 and 8, etc.) it is important to have the fine tuning optimized for the best results on the correct position. It is perfectly normal for some sets to have a herringbone pattern on the picture when, say, channel 2 is examined on the channel 3 position of the selector. In such a case, the fine tuning is most probably incorrect.

A herringbone pattern may, in some few cases, stem from internal causes. If, after exhausting the various possibilities listed above, the problem still persists, a substitution of one or more tubes in the video detector and video amplifier section of the receiver should be tried. This includes V6A in Fig. 6-2, V8A in Fig. 6-3, V5A in Fig. 6-4, V5A in Fig. 6-5, etc. In those cases where the video detector is a semiconductor diode, you may have to call in a professional servicer to do the job because careful soldering is required.

Fig. 8-6. Ghost lines or Barkhausen interference appear vertically toward the left side of the screen.

This also applies to those cases where tube substitution is of no help, suggesting that a defective component other than tubes may be responsible; however, semiconductor video detectors are seldom at fault.

BARKHAUSEN INTERFERENCE

Barkhausen interference appears as one or more ragged vertical lines, usually in the left portion of the picture. Sometimes Barkhausen interference may appear as a small group of adjacent lines somewhat as in Fig. 8-6. Such interference is most prominent between scenes (station breaks, etc.) or when the received station is one of the more distant (weaker) ones. This interference is invariably caused by a peculiarity (not really a defect) of the horizontal amplifier (V11 in Fig. 6-2, V10 in Fig. 6-3, etc.) tube. No tube tester will discern anything abnormal in such a tube, and nothing short of substitution will prove whether or not the problem is within the tube. Of course, it is just barely possible that the second tube may behave in the same manner, but this is highly unlikely. There are one or two other possibilities for which the corrective steps are simple and straight forward.

Check the tube caps on the horizontal amplifier and the HV rectifier (V11 and V12 in Fig. 6-2, etc.). They should be well seated and otherwise not intermittently connected. Check the door or cover of the cage containing the high-voltage components. It should not be loose or open. Check for the rare possibility that the TV lead-in is somehow bunched, possibly because of surplus length, near this cage or near the horizontal amplifier tube or general vicinity. Substitute a new tube for the horizontal amplifier.

There is one other and less preferable method for remedying this effect without replacing the horizontal amplifier tube. It involves the purchase of a little device, sometimes called a Barkhousen suppressor, and consists of a small magnet with an attachment (spring) band. This gadget is slipped over the horizontal amplifier tube's glass envelope and its position adjusted by rotating and sliding it up or down, while observing the Barkhouser lines on the screen. At the top cap of the horizontal amplifier tube there is a potential of thousands of volts. It is not necessary to come in contact with it since the magnet is slipped over the tube with the cap off and the receiver off. But care is the watchword here, nevertheless.

PICTURE GHOSTS

This malfunction is another one seldom caused internally, but it is quite easy to correct. A picture ghost is a duplicate picture, usually

Fig. 8-7. A ghost image, caused by multipath reception of two signals (one delayed slightly), appears to the left of the desired image.

weaker than the original and usually displaced somewhat to the right. Figure 8-7 is one example of this type of ghost. We are employing the word usually because there is also an unusual version of the same thing. A brief explanation will make it much easier to correct the fault.

It is common knowledge that a TV signal travels from the transmitting antenna to the receiving antenna. It is also a fact, although not so commonly known, that the TV image is reproduced on the home screen from left to right; that is, the picture is generated much in the same sequence as the writing (or reading) of a line on a page. It follows that portions of the TV picture on the right appear later than the portions on the left. Of course, the difference in time is extremely small (millionths of a second) and the eye sees the whole picture simultaneously; however, should two signals from the same TV station arrive one behind the other because one comes in a roundabout way the second signal will arrive a little later. This is known as multipath reception; that is, one signal travels the shortest path between the transmitter and receiver antennas, while the other follows a somewhat zig-zag path by striking some object on the way, then bouncing off, etc. This latter path is obviously longer, and takes more time. Figure 8-8 is a simple illustration of this occurrence.

If the TV receiver is far enough away from the transmitting center (such as a central metropolitan location) so that all stations lie in approximately the same line of path, the surest way to avoid or eliminate the type of ghost just described is to reorient the antenna. It probably has shifted so that it no longer points directly at the transmitters. It should be remembered that TV transmitters radiate in all directions (omnidirectional), while all receiving antennas are relatively sharply unidirectional; therefore, it is possible that the receiving antenna now is aimed at a tall building or a hill which acts as a reflector for transmitted signals. In that case the antenna is picking up a direct and a reflected signal.

If the transmitting location is not in sight, a map and compass should be used for correct pointing, although an aim-and-try (rotate antenna, observe picture, repeat if necessary) procedure is quite satisfactory. It is also fairly safe (although not absolutely certain) to assume that the correct direction is that in which most TV antennas in the immediate vicinity are pointing.

In cases where different TV stations are not in line but are in definitely different compass bearings, an antenna rotor is essential. This will permit the receiving antenna to be pointed directly at the transmitter without any compromise.

Fig. 8-8. Multiple signal paths between transmitter and receiver cause ghost images.

As more and more TV receiving antennas are made to receive a combination of VHF (channels 2 through 13) and UHF (channels 14 through 83) signals, an additional adjustment may be required for elimination of ghosts and for maximum signal, especially on the UHF stations. This adjustment consists of the raising or lowering of the antenna, while simultaneously rotating it, for minimum ghosts and maximum signal. Contrary to most cases of VHF antenna installation, the highest possible location is not necessarily the best for UHF. In fact, lowering a VHF-UHF antenna system as little as a few inches may make the difference between a poor UHF signal and a very strong signal. Since the VHF is not so critical it is safe to adjust the height for best UHF performance only. The direction of the antenna is fairly critical for all stations so this should not be neglected.

Tubes may cause picture ghosts, but such faults are confined almost exclusively to the front end of the TV set, specifically in the antenna stage (V1 in all block diagrams or V1, V1A in VHF-UHF TV sets). In those TV sets where the UHF tuner uses no tubes, the problem is virtually eliminated automatically, as the transistorized antenna circuit is seldom susceptible to this problem.

LOSS OF FINE DETAIL

This defect may be either component deterioration or due to misadjustment. Any portion of the picture where fine variations are present, such as a person's eyes or hair, a fine pattern on paper or clothing, and even the fine scan lines (raster) will be degraded by this defect. There are a few probable causes.

IF Alignment Incorrect

If the TV used to have good picture detail, a loss due to IF circuits misalignment will seldom occur, except and unless those circuits have been improperly adjusted. The remedy is obvious—complete IF alignment by a professional.

Fine-Tuning Control

Especially in very old TV sets this can materially degrade picture detail since the ultimate effect is similar to that of a misalignment. The remedy here, too, is obvious, except that the setting of the fine-tuning control also affects the sound and, as discussed previously, ghosts. An optimum, possibly compromise setting is recommended.

Defective Tube

Either the video amplifier or the video detector may be responsible. Checking the suspected tube is of little value here and substitution is recommended. As in the case of ghost problems, should a semiconductor diode be involved, or should tube replacement fail to solve the problem, the services of a professional will be required.

Improper Focusing

Although this misadjustment should be fairly obvious the beginner may not be aware of it at all times, since, in mild cases, defocusing can be tolerated, especially in action scenes (which applies to most TV programs). Picture focus deterioration with long use is not unusual and it should be checked whenever picture detail seems deficient.

As detailed earlier, focusing is best done by closely observing the TV screen, using a mirror if necessary, while adjusting the focus control for the sharpest horizontal lines on the screen. In those older receivers, changes in the ion trap magnet position should be made with circumspection as it is easy to degrade picture brightness, picture size and coverage of the screen edges by gross changes in the ion magnet position. Very slight and gradual rotation and sliding will easily show whether the focus is affected.

SMEAR

This defect, smear, is best observed on the larger areas of uniform illumination, such as a wall, the border between a person's white shirt and dark suit, etc. Practically every cause listed previously for loss of fine detail applies here, with two exceptions. First, focus misadjustment is not nearly as evident on large areas of uniform illumination. Second, there is a good likelihood that a defective component in the video amplifier circuits is responsible for the loss of good quality in the heavy areas.

One other potential cause for loss of heavy detail is misadjustment of the AGC. Since the AGC function sets the amplification level of the TV set, it is, in a manner of speaking, a coarse presetting for the contrast control, while the latter is, comparatively speaking, a fine control. As stated earlier the AGC should be set so that the contrast control is set about midway while receiving an average signal. Any setting of the AGC requiring the use of either the maximum or minimum extreme setting of the contrast control for an average signal is also likely to cause a type of distortion which ultimately evidences itself in picture smear.

LINE PAIRING

This is a picture quality defect causing severe image deterioration both in the fine detail and in the large solid areas, but particularly in the fine detail area. It is due to an accidental overlapping of adjacent raster lines. In appearance the number of horizontal lines on the screen is effectively cut in half, while the thickness of each new line and the spacing between them is doubled.

In Chapter 1 it was established that the visible rectangle (raster) on the screen consists of approximately 480 interlaced lines; that is, lines "painted" by the electron beam, first the odd-numbered set (lines 1, 3, 5, 7, etc.) then the even-numbered lines (2, 4, 6, 8, etc.). During normal operation the sync or precise timing pulses from the transmitter keep the spacing between the lines exactly right, so that, for example, line 2 falls exactly midway between lines 1 and 3, line 4 fits exactly between lines 3 and 5 and so on. In this manner any picture information on one of these lines is clear and distinct from any information on an adjacent line. Should, however, due to a malfunction, two adjacent lines overlap, two different sets of picture detail will overlap and neither will be clear. The term *pairing* of lines refers to just such an overlap. A failure in the sync circuits, due to either internal (TV set) or external (freak reception or transmitter malfunction) causes will often produce this phenomenon. To verify whether or not the TV set is responsible a simple procedure is involved.

Switch to a different channel, preferably three or four away. If the malfunction persists the fault probably lies within the set. Since it is extremely unlikely that two or more TV stations are having technical difficulties at the same moment, this source can be discounted. Similarly, freak reception, with possible interference from a more distant station on the same channel will produce just this effect, but is most unlikely to occur on a substantially different frequency. Thus, if the pairing occurs on channel 2, switch to channel 5 or 7 and observe. If freak reception is responsible the other channels will most likely not be affected. If pairing persists, the TV set is at fault.

Replace the horizontal control tube (V10A in Fig. 6-2), the horizontal phase detector tube (V9A in Fig. 6-3), or the horizontal AFC tube (V7C in Fig. 6-5). Each of these tubes, although somewhat different in name and operation, performs a basic function in a unified effort of controlling the timing of the individual raster lines. A defect in any of them may produce pairing and similar sync malfunctions. In testing these tubes careful checking for leakage after a few minutes of being in the tester is important. Of course, substitution of a new known-to-be-good tube is always conclusive.

Remove the sync tube (V8A in Fig. 6-2, V8B in Fig. 6-4, etc). This tube can be equally responsible for pairing, except that a serious malfunction here would also show up in a loss or poor performance of vertical sync. The picture would tend to roll or slide up or down one or more frames at random times; however, the malfunction may not be very obvious here.

Check the vertical hold control. Although this control setting is not very critical in the presence of a strong signal, a fine touchup may sometimes cure a mild case of pairing. Adjust this control for best interlace; picture rolling will automatically be corrected.

Check the AGC pot setting. The AGC adjustment cannot be overemphasized. It may seem paradoxical to the beginner and even to some TV servicer that too strong a picture signal may go hand in hand with too weak a sync signal. Since the AGC adjustment serves to establish the strength of the picture signal or contrast, it is quite clear that in an effort to obtain a good strong picture where one did not exist (quite often for other reasons), the AGC setting can be advanced to the point where the sync pulses are reduced below a safe minimum. Adjustment of the control should be made until good sync performance (no more pairing) is restored, even if this produces a somewhat weaker picture, which can be corrected by the contrast control.

MISSING ADJUSTMENTS

In a generalized description of causes and effects of TV malfunctions and their remedies, provision must necessarily be made for examples of controls and adjustments found in a great variety of makes and models of sets. Frequently, therefore, the TV owner may find that his or her set has no such control or adjustment. One good example of this is the ion trap magnet. While it is found fairly generally in older sets, such is not the case in more recent models. The same may be said for such an adjustment as horizontal lock or focus on black-and-white receivers. There are two explanations to this. In case of a "missing" control or adjustment, it is fair to assume that it is not required; it should also be added that in such cases the problem will not arise in the first place, at least not from this cause. In the other case, an exactly identical control or adjustment may not exist, although there seems to the nonprofessional TV do-it-yourselfer to be one with a similar name. Various horizontal adjustments are in this category. I have therefore provided in the various block diagrams a variety of cases in which the ultimate functions are the same although the identifications are somewhat different, either due to small design differences or simply because of different terminology used by individual manufacturers. The sync blocks in the various diagrams are a good example of this practice. Furthermore, in suggesting possible causes for one particular defect, I refer to these seemingly different stages for examination, indicating that they are functionally the same. Finally, in reading the instruction booklet or examining the various descriptions in the service information for a particular TV set, the owner will be able to identify the exact function of the different component groups or blocks even though the name may be slightly different from some typical TV set described here.

SHORT OR FIRE

This type of failure, short or fire, in a TV set, although it may be relatively hazardous, is usually easy to diagnose, especially when the failure is catastrophic, that is, the set stops working completely. There are a number of possible failures of this type, each requiring a different procedure and remedy.

Fuse Blows

If the main fuse (the one which is accessible without removal of the back cover) fails, it may be due to either a momentary overload, a weakness of the particular fuse or an overload due to a defect in the TV set itself. It is permissible to replace such a blown fuse with another one of exactly the same value. A lower value will blow without provocation, while a higher value may not protect the TV set and may even cause a fire. Carefully watch the behavior of the set immediately after replacement and for a while afterward. If no further difficulties arise (TV set works normally, no smells of burning appear and no crackling or frying sounds are heard), the incident may be forgotten. The assumption is that the fuse or a sudden surge in the line voltage was responsible, and the TV set is not at fault.

Replacement Fuse Blows

If the second fuse blows, whether immediately or soon thereafter, it must be assumed that the set is at fault. No further fuse replacement should be attempted.

Circuit Breakers

What has been said about the main fuse of the TV set applies equally to the circuit breaker—a device which has replaced the fuse in many TV sets. In simple terms, the circuit breaker is a type of lifetime fuse. While the ordinary fuse blows during an overload and has to be replaced with a new one, the circuit breaker opens and disconnects the TV set from the AC line. To restore operation it is merely necessary to reset the circuit breaker by pushing a button on the outside of the device.

There are two common types of circuit breakers in use today. One type can't be reset as long as the overload persists, so that there is little danger of unknowingly causing the TV set to operate under damaging or even dangerous conditions. If a breaker can't be reset, as evidenced by the fact that the pushbutton does not respond to normal pressure, it is an indication that the overload is still there.

Burning Smell but Fuse OK

It will sometimes happen that a fuse will not blow when it should. Fuses are not very accurate devices; their behavior is not 100% predictable. In such a case damage to the TV set is possible, depending primarily on how long the overload exists before the fuse goes (the correct and safe procedure is to immediately switch the set off) and whether it's a transformer or transformerless type set.

In transformerless sets, a *local* overload may be sufficient to start a resistor to overheat or actually burn out. A burning resistor emits some intolerable odors. While the remedy is beyond the scope of the beginner, the damage is seldom of major proportionsm git is usually caused by the total failure of another component (a capacitor, for example).

In transformer-type sets, in addition to the same possibility as just outlined for a transformerless set, repeated blowing of the correct size fuse is very likely to be due to a defect (short circuit) in the power transformer. It too may overheat and smell for a very short time (seconds, not minutes) without being permanently damaged. This, too, is a repair job for the professional, and his or her advice on whether or not replacement is required must be taken.

As in the case of a burnout in a transformerless set, a failure here seldom takes other parts with it. In other words, a burned out transformer usually does not cause damage to other parts; however, as a matter of cause and effect, a transformer failure, in addition to being due to an inherent defect, may also be caused by an earlier failure of another part. A defective rectifier tube (V14 in Figs. 6-2 and 6-3), semiconductor diodes (Figs. 6-4 and 6-5) or a short-circuited, filter capacitor in the main power supply may be responsible, in which case, they too, will have to be replaced when the TV set is repaired.

In addition to a defective power transformer or shorted capacitors, there are many other potential causes of fuse blowing, even if less common than these two; however, since in all cases the services of a professional are required, it is only necessary here to classify all fuse failures into two categories. One is a random failure not due to any permanent defect. This one is identified by the fact that the correct replacement fuse does not blow. The second is a causative failure due to a permanent internal defect or damage, which should be repaired by a professional.

In connection with random fuse blowing it should be pointed out that it may sometimes be caused by injudicious placement of the TV set. Since most TV sets draw hundreds of watts, it is absolutely essential to allow the heat to dissipate. Pushing a TV cabinet tight against a wall or some other piece of furniture is sure to cause overheating. Regardless of whether or not the fuse blows, the heat buildup shortens the life of the tubes and other components, not to mention the poor performance caused by thermal instability.

Chapter 9
Troubleshooting the Color Section

As has been pointed out, color and black-and-white receivers are essentially the same up to the video amplifier. In other words, all circuits and functions from the tuner through the video detector plus sync, sweep, and sound are common to both monochrome and color TV reception; therefore, the troubleshooting procedures discussed so far apply equally to color and black-and-white receivers. In fact, color TV troubleshooting, except for problems in sections that deal solely with the color signals, is easier when the color functions are disabled.

Referring to the color TV block diagrams (Figs. 9-1 through 9-4) you'll see that in each case an adjustment is provided for setting the color level of the set. If the color control is turned to one extreme (counterclockwise) it shuts off the color completely, leaving a normal black-and-white picture on the screen. Turning the control in the opposite direction (clockwise) gradually increases the color signal up to full color (and beyond, into distortion).

The color control is called different names in various receivers. In Fig. 9-1 it is called simply *color*. In Fig. 9-2 it is called *chroma gain*, while in Figs. 9-3 and 9-4 it is called *color control* and *color level*, respectively. In each case, this adjustment should be turned to the extreme no-color end of rotation before starting any of the troubleshooting procedures outlined previously.

NO COLOR BUT B&W NORMAL

The first and most obvious reason for the absence of a color picture is the setting of the chroma or color input control. Since this affects all three primary colors all color will disappear if the color control is accidentally turned off. A simple check of the control setting will determine whether or not it is the cause of no color. If it is, the control is simply reset (assuming no other adjustments have been changed) by eye while observing the screen.

Fine Tuning & Color

In addition to the obvious matter of setting the color level from no color to any desired intensity by the chroma or color control, there is another factor affecting the presence and intensity of color—the fine-tuning control. As discussed earlier (Chapter 4), reception of color depends also on the burst signal from the station and the 3.58 MHz color oscillator (block J, Fig. 4-1) in the receiver. Since the setting of the fine-tuning control naturally affects the received signal, it is possible, especially in a weak station, to have the fine-tuning control off its optimum position. It appears that while a strong enough picture is still received, the color will be absent or very faint. Adjustment of the fine-tuning control will bring the color component information up to an acceptable level. Since in all but very few TV receivers fine tuning is preset for each individual station, such an adjustment will not affect any other station.

Chroma Control Ineffective

If after advancing the chroma control there is still no color on the screen the failure probably lies beyond this point (between the control and picture tube). Furthermore, since none of the colors are present the defect must be in a circuit common to all colors, including the bandpass amplifiers (V15A in Figs. 9-1 and 9-2, V15A-V15B in Figs. 9-3 and 9-4), the 3.58 MHz oscillator (V15B in Fig. 9-1, V17B in Fig. 9-2, V20B in Fig. 9-3 and V18A in Fig. 9-4), the chroma demodulators and the final amplifiers. To find out which stage is defective begin by substituting tubes.

Bandpass Amplifiers—In some cases, the tube location diagram inside the TV cabinet may indicate two separate stages using a common dual-purpose tube. This is illustrated in Figs. 9-1 through 9-4, and is reflecting the fact that many tubes in modern TV sets are multiple-function or multiple-section types. Two, three or even more tubes in one envelope is not uncommon; so, when you remove a tube of this type, another function may be disabled, too. For example, in our typical block diagram in Fig. 9-1, half of V15 is the bandpass amplifier, while the other half is the subcarrier

Fig. 9-1. Block diagram of a typical color set. Semiconductor diodes serve as color demodulators.

amplifier; however, since the whole tube must be discarded if one half is defective, it does not harm to disable two functions at once. When a replacement tube is substituted, both functions will return to normal if the original tube was at fault. If the trouble is still there, then this particular (dual) tube is not the cause.

3.58 MHz Oscillator—As was mentioned in the general discussion of the color receiver process in Chapter 4, this function is essential for the recovery of color. Putting it another way for the purpose of troubleshooting, if this oscillator fails all color will be absent. Replacement with a good tube will prove or disprove this suspicion. As in the case of the chroma amplifier, a twin tube may be involved here, but this does not affect this trial-and-error check on the subcarrier oscillator.

3.58 MHz Crystal—A brief explanation should be helpful. As mentioned earlier in the general description of the TV system, certain very precise timing (or sync) signals are sent from the transmitter for cueing or timing the various signal sequences on the receiver. One of those signals exclusive to the color TV set is the subcarrier *burst* signal. This is a precise timing pulse which keeps a locally generated signal at its exact timing, and it is commonly called the 3.58 MHz reference. Without this reference signal no color reproduction can take place. The heart of this oscillator is an electromechanical element called a crystal. In some TV sets it is a plug-in device and resembles the drawing in Fig. 4-2.

To remove the 3.58 MHz crystal, grasp it and pull upward, after first noting carefully the appearance of the picture tube screen. Often, a very poor crystal may still allow some color to appear on the screen. If the removal of the crystal makes no difference, then the fault does not lie here; however, if the slight coloration previously seen does disappear, the crystal is probably defective (or intermittent) and should be replaced. Classically, when an inoperative 3.58 MHz crystal is encountered the screen will appear to have only blue and green—no red. Also the tint or hue control will have no effect on the color present on the screen.

This is a standard item found at most radio supply houses catering to the TV servicers; therefore, a replacement crystal of the same type is readily obtainable. Where the service data lists a crystal by a particular manufacturer's part number, it may be more practical to obtain a replacement from the service department of that manufacturer.

Fig. 9-2. Block diagram of a color receiver using a dual-purpose tube as the color demodulator.

In most recent TV sets the 3.58MHz crystal is a pigtail type; that is, it is not like the plug-in type just described but is intended to be soldered in the circuit similar to a resistor or capacitor. In such a case, the crystal is "worked around" in the troubleshooting procedure until all the other alternatives are exhausted, and if none solves the problem of the absence of color, it is assumed that the crystal is at fault and a professional servicer is called (See Chapter 10 on do-it-yourselfer procedures on soldering printed circuit components as an alternative to calling a TV servicer).

Color Killer Tube—There is just a possibility, although not very probable, that a malfunction in the color killer tube would cause the color absence. As explained in Chapter 4, the function of the color killer stage is to shut off the chroma amplifiers in the absence of a color signal. At first glance this may not sound very logical; why shut off the color amplifiers when there is no color to amplify? Actually, however, this is logical as can be easily shown.

On all color block diagrams, Figs. 9-1 through 9-4, each of the three color guns of the picture tube receives a signal from a color amplifier. The signals are electric currents or voltages, not actual colors. If any of the color amplifiers feed an electrical signal into one of these guns, that gun will produce a color on the screen, the particular color depending on which gun is being fed. To insure, therefore, that no color is shown when a black-and-white picture is transmitted, the color killer keeps the color amplifiers shut off until a color burst signal (present only when a color picture is transmitted) commands the color killer to let the amplifiers perform normally. If, due to a malfunction, the color killer tube keeps the chroma amplifiers permanently shut off, no color will come through. Replacement with a good tube will prove or disprove this suspicion.

Color Killer Setting—In all cases where there is an adjustable control marked *color killer adjust* or just plain *killer*, the setting of this control should be checked, even before suspecting the color killer tube. It is just possible that this control is set so that the bandpass amplifier could never pass a signal. Incidentally, the color killer control is normally set so that all traces of color just barely disappear on a blank channel; then advance it ever so slightly beyond this point. When some professional servicers set the killer they use an indicating instrument of one type or another, but the suggested procedure for a beginner is effectively equivalent to that above.

Component Failure—If all the preceding procedures fail to restore the color to an otherwise normal TV set, the fault must lie in

Fig. 9-3. In the color set represented by this block diagram, the color demodulators drive the picture tube directly.

one of the internal components in the path of the color signal, such as a coupling capacitor or a failure of a supply voltage to one of the tubes. In the first case, the tubes function normally, but the signal path is interrupted. In the second case, one or more tubes are inoperative due to the failure of the correct operating voltage. Whichever the case, the services of a professional servicer are required.

WEAK COLORS

If all the colors are more or less uniformly weak, that is, no one color is much worse than the other two, the fault again lies in a common color circuit. The troubleshooting sequence here is logical.

Bandpass Amplifier

A weak tube in either of the bandpass stages, if there are two, will most likely decrease the color contrast.

Color Control Setting

If the color control has not intentionally or accidentally been backed off, in other words, if this control is in its usual position or setting, try advancing it clockwise. Advancing the color control will probably restore the color to a good level; however, if it does, it is an indication of weak tubes, especially if advancing the control to its maximum position still does not produce a normal color picture. Incidentally, it should not be necessary to have this control beyond its approximate mid-position if the bandpass amplifiers are in normal working condition.

Color Killer Setting

An improper setting of the color killer control can produce the same symptoms as weak bandpass amplifiers. Unless one is sure that the setting of this control is correct from previous observation under normal set operation, the possibility exists that the setting of this control is responsible for reduced color performance. A quick check on the proper setting is to switch to a station transmitting no color at the moment and adjust this control as outlined previously for the normal setting to obtain a clean black-and-white picture.

Selective Color Troubleshooting

Up to now we have discussed what might be called total or catastrophic color failures. These involve either all of the color

Fig. 9-4. In this block diagram one multifunction tube demodulates the color signals and drives the picture tube.

portions of the receiver or a vital link in the chain which, if broken, stops any color from getting through. In the following pages, we shall consider local color defects, when a partial break in a secondary chain is responsible. For this purpose it is necessary for you to be able to clearly identify the function and tube or tubes involved in each case, although you will not have to understand the technical version of how and why. Such identification is a bit confusing because different terms are used by different manufacturers for the functions and because of differences in functions between different TV sets; however, a study of the block diagrams of color receivers, beginning with Fig. 4-1 and continuing with Figs. 9-1 through 9-4, will enable you to identify the basic functions in each diagram, regardless of the individual differences. Since Fig. 4-1 is a *typical*, not necessarily *actual*, receiver its various blocks show *functions* as distinguished from actual examples. By contrast, all the other diagrams just listed represent *actual* receivers. Furthermore, there is considerable variation between them to afford a good sampling of different methods of achieving the same end result.

COLOR FUNCTIONS

After the color detection takes place the three primary colors exist individually, up to and including the picture tube; thus, when it is stated that one color is absent, it is understood that the various shadings or combinations of this color are also missing. A missing red signal denotes a picture made up of the remaining two colors only. This implies that the absence of blue would show no shades of blue and also an abnormal green, since the green signal is largely dependent on the existence of a normal red and blue. This may not follow from the logical understanding of colors in nature because it is based, instead, on the technical structure of the color system in the TV set as we shall presently show.

An examination of the various block diagrams in Figs. 9-1 through 9-4 will disclose some important facts. There is a group of function blocks called demodulators and an adjacent group called amplifiers. Further observation will disclose that the demodulators are identified as the X-demodulator and the Z-demodulator, as in Fig. 9-2, and consist of two tubes or two sections of a twin tube. In the latter application they are marked to indicate that they are one-half of a two-in-one tube (for example, V18A is one-half of a 6DT8). In other cases, as in Fig. 9-4, they are part of a three-in-one tube, such as a 6LE8. In one case (Fig. 9-1), no tubes are used; instead four semiconductor diodes (D1, D2, D3, and D4) are employed. In Fig. 9-3 the terms *R-Y* and *B-Y* are used instead of *X* and *Z*. In the next group of function blocks, in practically all cases, they are marked R-Y amplifier, B-Y amplifier and G-Y amplifier.

In some cases (Figs. 9-3 and 9-4) there are no such amplifiers at all. In one example (Fig. 9-4) one multipurpose tube (V16, 6LE8) performs all the functions of the demodulators and final color amplifiers. In the second group of functions, those of final color amplifiers, there are also a number of different arrangements.

Returning to the names and functions of the various stages, the demodulators serve to extract or separate the color signals from the combined electrical signals which served to carry the color information through the TV set. After the demodulators the signals are the ultimate color information, which in some TV sets requires only additional amplification before application to the three color guns of the picture tube. As to the color identification, both the demodulators and the following amplifiers carry the designations R-Y, B-Y and G-Y.

Notice that the hyphen between the letters are intended as minus signs and should read G minus Y, R minus Y, etc. These names describe a method of color processing technically called *difference color system,* that is, difference color demodulators and difference color amplifiers. In the beginning at least it is not necessary to know the technical reasons for this. Instead, it is perfectly satisfactory to assume that R-Y stands for red, B-Y for blue, and G-Y for green. What needs to be made clear here is that in all TV sets there will be no G-Y demodulator because this is achieved as part of the R-Y and B-Y functions. Furthermore, in some TV sets the demodulators are marked X-demodulator and Z-demodulator. The first of these is equivalent to the R-Y demodulator and the second to the B-Y demodulator.

INDIVIDUAL COLOR ADJUSTMENTS

In addition to the previously mentioned overall color gain control marked *chroma control* or *color*, there are two other sets of adjustments in the color-only portion of the receiver which must be taken into account when partial color deficiency is evident. One set of controls is called *drive*, the other set is called *screen*. There is usually one in each set for each color, namely, red drive, blue drive and green drive and red screen, blue screen and green screen. Sometimes the red drive is omitted.

Drive & Screen Controls

Drive and screen adjustments are important to the overall representation of a color picture. Unless these *setup* controls are properly adjusted it will be impossible to achieve a good color representation of the color telecast. Although these two sets of controls seem to have a similar effect on the color quality, their purpose is in two different extremes. Drive controls govern the highlights while the screen controls set the operating point for each gun. That is, a misadjusted screen control will have a more dramatic effect on the picture than a misadjusted drive control.

Improper adjustment of the drive controls may produce, depending on whether the adjustment is too little or too much, either too faint a picture in the dimmer portions of the scene or too strong on the brightest portions of the scene. By comparison the screen adjustment can make the whole picture look either too low or weak (insufficient) or too high (oversaturated). The normal adjustment for these twin-type settings is, first, to get the complete range of variation from dimmest to brightest, and second, to get this complete range up to a level which is an acceptable picture.

Gray-Scale Adjustment

The previous method for adjusting these two sets of related controls is based on purely subjective criteria—you adjust the controls until, to the best of your knowledge and observation, the color picture is most acceptable. A somewhat more impartial method, and one that the professional servicer usually follows, is what is known as the gray-scale adjustment. This simply means that the three guns of the picture tube are balanced to produce as nearly a perfect black-and-white picture as the design of the TV set allows.

As explained earlier in connection with the characteristics of light, white light is a balanced combination of all colors. In a color TV system this is modified to read: a normal black-and-white picture is the result of the correct combination of the outputs of all three color guns. It is the purpose of the drive and screen adjustments to produce such a correct combination. The term gray-scale adjustment is borrowed from photographic terminology and denotes a continuous range from pitch black through all shades of gray to full white. A normal black-and-white photograph should show all these levels, provided, of course, that they existed in the original scene. In a three-gun color picture tube proper setup of the gray scale is a prerequisite for a balanced color picture. But even in a gray-scale adjustment carried out by a professional servicer there is a certain amount of *magic* involved, except in those cases of more recent sets where a detailed adjustment procedure, applicable to a particular set only, is provided by the manufacturer. In such cases there is often an auxiliary device marked *normal-service switch* or similar terminology, indicating that the switch is left in the normal position for use of the set (some normal functions are disabled in the service position), and temporarily switched to service for adjustment purposes. But even in these cases final touchup is still done by the servicer by magic. Based on all these factors, the best procedure for the beginner still is to use careful observation and judgement, proceeding as outlined in the steps to follow.

1. Turn the receiver on and let it warm up for about 30 minutes.
2. Turn the color or chroma control to minimum. If the set is equipped with a control that varies the gray-scale adjustments, set it to midrange. Some brands of receivers call this control picture quality or tint. Do not confuse this control with what is called the tint control on some makes. In one case we are discussing the hue control and in the other it is the raster coloration control. If it varies the coloration of the raster with the color control at minimum, it is the control in question. Set it to midrange so that when this procedure is completed you will have the desired and intended range of the control.
3. Adjust the front-panel controls for normal viewing with the exception of the foregoing controls. Tune in a medium-strength station and fine tune the tuner.
4. Vary the brightness control from minimum to maximum, observing the entire screen for discolorations or tinting. If there is a normal black-and-white picture from low brightness to high brightness, it is not necessary to go any further. If this is not the case, proceed to step 5.
5. If the TV set is equipped with a normal-service switch as described earlier, set it to the service position. If the receiver isn't equipped with this switch the procedure will be slightly more difficult, but it can be accomplished with a little extra care. With the switch set to the service position a horizontal white line from side to side will appear.
6. Set all of the screen controls to minimum, noting their original position.

7. One at a time adjust the red, blue and green screen controls so that each color just begins to appear on the screen. Adjust all three so that each gun is turned on at approximately the same level. This is best accomplished in a darkened room because red will not appear so bright as the other colors and can be more readily seen in this manner. In receivers not incorporating a normal-service switch, you must eyeball the screen at a low-brightness setting to determine when each color just appears on the screen. The screen controls are set so they have the same threshold level for each gun.
8. If the receiver is equipped with the normal-service switch set it to the normal position. If it is the type without the switch set the brightness control to a normal viewing level.
9. One at a time, bring each of the drive controls down from a maximum position to a point that results in a true black-and-white picture. Most often the red drive is not adjustable and the end result will show the blue at maximum and the green slightly reduced.
10. Check the picture for gray-scale tracking at all brightness levels. When you are satisfied the job is as good as can be obtained, check the color presentation by turning up the color control.

COLOR INTERDEPENDENCE

Although separate final amplifiers may be used for each of the three primary colors, there is a certain amount of interdependence between them. So much so that a failure in one color will probably affect them all because of the technical nature of the color image structure. As detailed earlier most TV sets have but two color demodulators, a R-Y (or X) and a B-Y (Z). The G-Y does not exist separately because it is in a sense a composite of the other two as far as extraction of the signal is concerned.

We do not, of course, mean that a simple mixture of red and blue will produce green. What is meant is that the color transmission standards are such that proper blending of the R-Y and B-Y signals will also result in a G-Y output for feeding the G-Y final amplifiers. Furthermore, just as R-Y and B-Y (red and blue) components combine to produce the G-Y (green) component, a defect or malfunction in one of the first two will produce a corresponding deficiency or abnormality in the third one. For example, loss of the red signal will, primarily, remove virtually all the red from the picture and secondarily, distort the remaining color and particularly the green, into some other hue or combinations of hues. A detailed troubleshooting sequence for the absence of a particular color follows.

RED MISSING

There are three areas which affect the red color in the picture—the demodulators, the individual color amplifiers and the picture tube. Based on some possibility of side effects on the other primary colors as well as on the inbetween shades the test sequence is simple.

Blue & Green appear Fairly Normal

Since you probably do not have access to a color-bar generator (except in the case of a built-in color generator), the judgment of color must be done subjectively. Look for known blue picture elements, such as sky, or known greens, such as grass, to judge. If such areas of known color seem to show normal color the trouble is probably not in the demodulators.

Replace the R-Y Final Amplifier

This is V16 in Fig. 9-1, V19 in Fig. 9-2, etc. If normal color returns, you've found the cause of the trouble. If there's no substantial improvement, replace the original tube.

Locate the Red Screen Adjustment

This control affects only the red gun of the picture tube. Before manipulating this control, mark the starting position (knurled or slotted shaft) with a pencil or a piece of masking tape. Advance this control gradually, usually clockwise, to increase the red color. If no noticeable effect is observed, restore the shaft to the starting position. If gradual improvement results from turning the control advance the adjustment further until red is normal.

A note of caution is in order here. First, if the red screen control had accidentally been backed off (this is a possibility, but not a probability), advancing it to normal color production is perfectly proper; however, if the control has to be turned all the way to barely obtain a fairly good red, a defect elsewhere in the system is responsible and compensating for this fault with the screen control is not advisable for the good of the picture tube. While it is quite possible that the red gun in the picture tube has deteriorated, this is not too common since in normal use all three guns deteriorate more or less equally.

Replace the R-Y Demodulator

This is not much of a possibility in view of the fact that the green seems normal; it would not be normal if the R-Y (or the X) demodulator were defective. If this replacement does not cure the red deficiency when advancing the red screen is only partly effective, a defect in one of the components (a capacitor or resistor) is most likely the cause. Incidentally, the red screen adjustment is helpful if one of the color guns is defective and a new picture tube will ultimately be required. Under these circumstances no harm is done in advancing the control and obtaining as much use of the tube as possible before a replacement must be purchased.

Touchup the Red Drive Control

This is important because the drive and screen controls for any one color are interdependent and may seem to act alike, especially in the hands of the inexperienced. As in the case of the screen adjustment outlined previously, it is good practice to observe the initial setting of the drive adjustment so as to be able to return it to its original setting if no change is apparent. The original setting of the drive control is important since, unless there is evidence to the contrary, the control was set for the proper operating conditions for the particular gun (red in this case); however, you're safe in making a slight adjustment if there is clear evidence that such an adjustment, combined with a corresponding adjustment of the screen control, improves the red color sufficiently to make it acceptable. If sucl adjustment fails to show a substantial improvement in the red color a defective component is probably responsible.

BLUE & GREEN MISSING

As in the case of red the blue channel has its own demodulator (B-Y or Z-demodulator) and its final amplifier (B-Y amplifier). If the screen shows a color picture containing mixtures of red but not blue or green the indications are that a defect common to both blue and green is responsible, since blue is one component of the green signal. The first and most likely suspect is the B-Y demodulator (or the Z-demodulator in some TV sets). The first step is simple.

Replace the B-Y Demodulator

If the blue color appears, the green will also be present. If replacement of the demodulator tube does not help there is one other tube suspect; however, since both blue and green are missing, both the B-Y and the G-Y amplifiers would have to be defective. This would be rather unlikely if the B-Y and G-Y amplifiers were two separate and distinct tubes. A close look at the typical block diagrams in Figs. 9-1 through 9-4 will show that in almost every case the two amplifiers are two sections of a multisection tube. A failure in both sections, therefore, is no longer unusual. Consequently, the next step is to check these tubes.

Replace the B-Y & G-Y Amplifiers

Replace the two-in-one or three-in-one tube comprising the B-Y and G-Y amplifiers. This is V16 in Fig. 9-1, V19 in Fig. 9-2, etc. If the fault lies in the color amplifier this step should be effective. If it does not restore the blue and green colors try the next step.

Reduce the Chroma Gain

Reduce the chroma or color control to minimum so that no color is present. If the black-and-white picture is normal the picture tube is okay and the fault lies in a defective component, an open capacitor or resistor. Of course, there is just the possibility that both the blue and green guns of the picture tube are defective, but you'll need a professional servicer for positive diagnosis and repair.

BLUE MISSING

This malfunction is evident by a predominantly red and green picture with an almost total absence of blue. Since the green is present the blue signal must exist in the output of the B-Y (or Z) demodulator. The only possibilities for a loss of the blue color are indicated in the following paragraphs.

Defective B-Y Amplifier

Replace the B-Y tube whether it's a separate tube or one section of a multifunction tube. A failure here will be cured by the substitution of a good tube. If no satisfactory improvement is obtained here try the next step.

Blue Screen Adjustment

Adjust the blue screen control on the back of the set. As discussed earlier a partial deterioration of any one color gun in the picture tube can be (temporarily, at least) compensated by advancing the screen control. If, however, only a slight improvement (or none) is realized from a major adjustment of the screen control the fault lies either in a defective component in the blue signal path or, less likely but still possible, the blue section of the picture tube is gone. In either case, outside help will be required.

On the other hand should the blue screen adjustment produce a substantial improvement a further correction may be attempted by a small adjustment of the blue drive control. Since this points to a deterioration of the blue gun these two adjustments can restore the tube to a reasonably normal operating condition for as long as the tube will hold out. Again, this precaution is worth repeating: no decision to replace a picture tube should be made without the concurrence of the professional TV servicer.

COLOR PRESENTATION UNSTABLE

This could be either tearing of the color picture, similar to the diagonal zig-zag tearing of a black-and-white picture when the horizontal frequency adjustments are off (see Fig. 2-6), or in less severe cases a partial breakup with the color wavering and general picture instability, often only intermittently.

It is assumed that the do-it-yourselfer is following the general admonition given earlier with regard to any color troubleshooting-before attempting any work in the color portion of the receiver, the performance in monochrome is checked. If the symptoms are the same as in color operation the fault lies not in the color-only portion of the set, but in the common section, and repair should be made there. Only if the monochrome performance is normal should work on the color section be attempted.

As a general rule this type of behavior suggests improper operation of one of the color sync stages. This includes, depending on the particular TV set, the burst amplifier (V15B in Fig. 9-2, V18 in Fig. 9-3, etc.), the burst gate (V7C in Fig. 9-1), the 3.58 MHz oscillator (V17B in Fig. 9-2, V20B in Fig. 9-3, etc.) and the 3.58 MHz crystal itself. In some TV sets the sync and oscillator functions may be called color sync, burst gate, subcarrier amplifier and chroma oscillator control. In all cases the functions are the same even if the names differ. These functions are the color sequencing and positioning precisely in step with the pictures being transmitted. Any malfunction here will throw the received picture out of kilter. The procedure in troubleshooting these malfunctions is the next step.

Replace Color Sync Tubes

Since the number of sync-related functions and tubes used in each case differ widely, it is necessary to determine the sync lineup in each particular set. Using typical block diagrams as examples the sync stages in Fig. 9-1 consist of burst gate V7C, and subcarrier amplifier V15B in the order given. The corresponding sequence in Fig. 9-3 is gated burst amplifier V18, phase detector V19A, reactance control V20A and 3.58 MHz oscillator V20B. In Fig. 9-2 the sync lineup is burst amplifier V15B, reactance control V17A and reference oscillator V17B. In Fig. 9-4 it is burst amplifier V17A and 3.58 MHz oscillator V18A. These typical diagrams are based on actual TV sets and cover a large portion of all color sets in use today. Should one exceptional case have a somewhat different arrangement or designation you can identify each function by comparing the names and functions with those given here. The word *chroma* may be used for the word *color* and *control* is often used for *sync* or *burst*.

Although the replacement or substitution of tubes in the order given is the most direct way of finding a malfunctioning stage it does not mean that random substitution will not ultimately produce the same results. But for logical procedures the sequence should be followed, as it proceeds, in all cases from the color IF (bandpass) amplifier to the 3.58 MHz oscillator. A final general reminder: in all the sets represented by the block diagrams in this chapter the 3.58 MHz stage contains a crystal, and it is just as vulnerable a link in the sync chain as all the other tube stage enumerated here.

Replace First Tube in Sync Lineup

Very often the first tube in the first sync lineup will be one section of a multifunction tube, but this does not change the procedure. If only a minor improvement in stabilization results you have not found the trouble. In Fig. 9-1 the first tube is burst gate V7C, one-third of a 6M11 tube. In Fig. 9-2 it is burst amplifier V15B, one-half of a 6BN11. In Fig. 9-3 it is burst amplifier V18, a 6EJ7. In Fig. 9-4 burst amplifier V17A is one-half of a 6AF9.

Replace Second Color Sync Tube

In Fig. 9-1 this is sub-carrier amplifier V15B, one-half of a 6BW11. In Fig. 9-2 it is V17A, one-half of a 6GH8A, the reactance control. In Fig. 9-3 the second tube is V19A, the sync phase detector; in Fig. 9-4, it is 3.58 MHz oscillator V18A. If no better than a slight improvement is apparent the fault lies elsewhere and the procedure continues.

Try A New Crystal

Where the second tube involved the 3.58 MHz oscillator, as in Fig. 9-4, replace the 3.58 MHz **crystal** if it is a plug-in type. It is also advisable in this case to observe whether the color breakup is always present from the moment the TV set is switched on or if it appears later after the set warms up. If the latter is true, the crystal is a most likely suspect, as these units sometimes stop functioning (oscillating) after a period of normal operation but may return to normal if the set is switched off, then immediately switched on again.

Another possibility is poor contact at the crystal connecting pins. As described earlier in many TV sets the crystal connects to the TV set by means of two thin wire pins, similar to those on miniature tubes. Applying a gentle pressure on the crystal case may momentarily restore normal operation if poor contact is the cause. In other TV sets the crystal is soldered in, like any other part, and replacement is not so easy (see Chapter 10). As a final note it remains to be said that a crystal failure or a 3.58 MHz oscillator failure usually, even if not always, causes a total loss of color information, not just breakup.

A Final Measure

In TV sets similar to Figs. 9-1 and 9-3, replace the next likely offending tube. This is the oscillator control V20A, one-half of a 6GH8 in Fig. 9-3. Finally, in all types of TV set lineups the last stage is again the 3.58 MHz oscillator or the reference oscillator. If none of those steps seem to make an adequate improvement in color stability or normal picture the fault probably lies in one of a number of components in this general area, requiring troubleshooting by a professional using specialized instruments.

It should be added here that in a number of cases of mild color instability the fault lies outside the TV set. Occasionally, a color is transmitted under circumstances making perfect stability unrealizable. The owner can best verify this by switching to an adjacent color channel. If the problem disappears the probabilities are that the station, not the TV set, is at fault.

COLOR TV ANTENNAS

Commercially available TV antennas vary all the way from a piece of ribbon lead-in type of wire cut to some approximate length, costing a little over a dollar, through various dipole versions, all the way to multielement, multibay structures, costing hundreds of dollars. The more elaborate the antenna, the higher the *gain* (a term indicating the strength of the signal delivered to your set), and vice versa. An important characteristic closely related to gain is the bandwidth.

The second factor, although much less scientific, is nevertheless just as important: viewer attitude. It is a fact that the average observer is far more tolerant when viewing a black-and-white picture than he or she is of color. This is due, in part at least, to the fact that even a faint picture can be seen and followed in monochrome, while the weak picture in color is intolerable and actually much more difficult to follow. Taking both of these into proper consideration the following may be said about the color TV antenna requirements.

First, the satisfactory color antenna must have adequate gain. While almost any antenna, given a fair to poor signal, will provide an acceptable black-and-white picture, only a sufficiently high-gain antenna will serve as well for color reception under the same circumstances. Second, the color antenna must have adequate bandwidth. That is, it must be able to intercept and receive a strong enough signal over the whole expanse of the TV frequency spectrum, from channels 1 through 13 and possibly channels 14 through 83. Of course, this does not mean that a narrow-band antenna will receive only part of the picture. It will receive a single channel or possibly two or three. Narrow-band antennas are usually *cut* for specific frequencies.

While it is beyond the ability of the TV owner, no matter how well technically informed, to determine the characteristics of an antenna he or she wishes to purchase, there are a number of ways this information may be obtained without great effort. First, there are a few consumer organizations who do impartial testing and reporting on manufactured products, antennas among them. Their reports are published and may be easily obtained for a minimum cost, or they may be seen in most libraries. Second, and without even any factual data, and contrary to other examples, the cost of the antenna is a good indication of its capability. As a matter of confidence the purchaser may rely on the reputation of the manufacturer, be it a TV manufacturer who has antennas made under its label, or the antenna manufacturer itself. The only precaution one must observe is to steer clear of the sensational *amazing discovery* type of antenna bargains offered by (usually unknown) mail-order advertisers. The old cliche that good quality does not come cheap is very true here. In cases of color abnormalities, such as smear, color over-lap and of course color snow, assuming that the set has otherwise been checked out, it is advisable to consider the antenna as the possible source of the trouble.

The foregoing observations are not intended to suggest that one should always buy the most expensive antenna. Not only is this unnecessary, but sometimes actually inadvisable for the following reasons.

The purpose of any antenna is to provide an *adequate* signal for the particular receiver location, not the maximum signal possible. A simple perusal of catalogs and advertisements will show that antennas are classified according to intended use; thus, there are antennas for metropolitan, suburban, near-fringe, fringe and deep-fringe applications. Advertisements often specify mileage ranges for each type, but these may be taken with a grain of salt since they are for *ideal* conditions, not usually realized. A practical guide would be to purchase an antenna one grade better than indicated by the advertised mileage; thus, if the TV is located in the near-fringe distance from the station, a fringe antenna would be adequate. Of course, local conditions, such as locations behind a hill, etc., will adversely affect the signal. In such a case, it is not illogical to see what size and even type of antenna is used by others in the immediate vicinity. Having determined how big an antenna to get, it might further be advisable to get one of a reputable manufacturer.

COMBINATION VHF-UHF ANTENNAS

In deciding on the type of antenna consideration should be given to the reception requirements in the UHF portion of the TV band, both for the UHF reception as well as to the incidental effect on the more popular VHF stations. Since most network as well as independent TV stations of importance are concentrated in the VHF region, while UHF service at present is primarily of relatively minor scope, it behooves the TV owner to decide, before installing an antenna, where his or her primary interest lies. If it is in VHF only, then the best *VHF only* antenna will be best since almost all combination VHF-UHF antennas are but a compromise, with less than optimum reception in either range. If, however, the set owner is interested in the best reception of both types of stations, then a separate UHF antenna is desirable; even the very largest combination antennas give but mediocre to marginal UHF reception. The only exception to this situation is in case of suburban or somewhat more distant locations, or similar areas of high-signal strength. Here a combination antenna will be adequate for both VHF and UHF.

COLOR SMEAR

In a black-and-white picture a condition similar to smear often results from an improperly installed or wrongly oriented antenna. The usual name for such an appearance is ghosts—a slight displacement of the picture edge as if two superimposed identical pictures were slightly out of register or slightly displaced from one another, either to the left or to the right. One check on the accuracy of this diagnosis in a color picture is to turn the chroma (or color gain) control all the way off so that a black-and-white picture remains. If evidence of ghosts exists the diagnosis was probably correct. Adjusting (rotating) the antenna, if it is not otherwise defective (broken elements, poor lead-in connections at the antenna, loose elements, etc.) should eliminate ghosts or at least reduce them to a minimum and effect a marked reduction in the color smear.

DEGAUSSING

A fairly common cause of both color smear and color mixing or distortion, particularly around the edges of the picture, is accidental magnetization of some of the structural metal around the face of the picture tube by stray magnetic fields—AC lines in the wall, a heavy-current appliance in the immediate vicinity of the TV set, etc. No amount of adjustment can remove this color distortion. Only a demagnetization or degaussing of the picture tube will restore color purity.

While some of the more recent manufactured TV sets have a built-in degaussing capability most of the earlier color sets are not so equipped. While the process of external degaussing is relatively simple and quick it does require a degaussing device, a large (6- to 10-inch diameter) ring made of many turns of insulated wire terminated in an AC plug. In operation the device is plugged into an AC outlet and moved about (in a definite manner and speed) in front of the TV picture tube. A few moments are required to demagnetize the steel frame and rim of the color tube. While it is not particularly difficult for a beginner to construct and use such a device, its very infrequent use makes the task unjustifiable. Actually, it may be more economical to have a professional do it for the nominal fee. Incidentally, inexpensive degaussing coils, sometimes advertised as *color purifiers*, have appeared on the market and come with complete simple instructions for their use.

Chapter 10
Solid-State Circuitry

Before listing some of the major differences between tubes and semiconductors it should be made clear that transistors and diodes perform the same functions as tubes; that is, they detect, amplify, oscillate, etc. The end result, whether picture or sound, is exactly the same, regardless of how it was achieved. Nevertheless, the structural and operational differences between tubes and transistors are very sharp. Ones that are important to the consumer are:

- Unlike tubes, transistors need no heaters for their operation. This is perhaps their greatest single advantage, resulting in greater stability, longer life and more economical power consumption.
- Transistors are low-voltage devices. Most operate on between 1.5 and 15 volts, although a few rare ones require as much as 100 volts or so. By contrast, tubes usually require a few hundred ohms.
- Transistors are not fragile. Shock and vibration are far more harmful to tubes than transistors. This relative immunity to vibration combined with low-temperature operation makes for long life.
- Transistors are simpler to manufacture, hence less expensive.

EFFECT ON TV DESIGN

Solid-state technology has produced a number of changes in the TV industry. To the do-it-yourself consumer, most of these changes have been positive. And although there may be a few disadvantages, as far repair by the owner is concerned, on the average the consumer is ahead. Some of the more obvious advantages are:

- Smaller size—With the proliferation of semiconductor use a major miniaturization technique has evolved, not only with respect to transistors versus tubes, but also for almost all other components except, of course, the picture tube and the speaker—capacitors, resistors, hardware, wiring, etc. A by-product of the parts miniaturization has been a decided improvement in the quality of components.
- Lower heat—Not only has the power consumption been greatly reduced, but heat generated also has gone down correspondingly. Less heat means less waste, but far more important, less wear and longer life for components. This is almost entirely due to the replacement of tubes with transistors, although some little progress in this direction was independent of the transistor factor.
- Less wiring—One of the greatest and most obvious effects of the solid-state technology has been the tremendous reduction in wires and wiring. And although wires are not usually a source of trouble, their elimination wherever possible is a distinct plus for economy of space, labor and cost.
- Modularization—With parts being much smaller, it has been logical to group a larger number of them on a single common subassembly; thus, printed circuit modules or boards came into vogue. We shall say more about these devices when describing do-it-yourself techniques.
- Repair standarization—When a TV set consisted of almost 100% discrete parts, no two jobs were alike, that is, the manner of repair, and to some extent the efficacy of such a repair, was largely dependent on the individual TV servicer, professional or do-it-yourselfer. The PC board and module have eliminated a lot of haphazard repairs. For any particular repair, the replacement module connected exactly the same way as the original—the repairman has no choice in this matter—will produce essentially the same normal operation of the set, regardless by whom it was repaired.

At first glance it might seem wasteful to replace a complete module when only one of its components is defective. Actually, this is not necessarily true. While it is sometimes possible to quickly locate the defective part and replace *it* only, thus saving the old board, in the majority of cases the replacement of the complete module is more economical. It saves time, money and possibly even avoids the risk of damage to other components in the immediate vicinity of the defective part—a most unpleasant experience in view of the fact that the damage may not be discovered until after reassembly of the TV set.

SERVICE TECHNIQUES

There are a few precautions as well as some new techniques to be observed in working with solid-state circuits.

- Care must be exercised that nothing is disturbed in the vicinity of the part being removed.
- An *absolute minimum* of heat must be used in soldering components.
- Physical force where necessary must be applied very sparingly. Miniature parts simply cannot stand rough handling, by hand or with tools.

The following few tools are recommended as an essential minimum in servicing solid-state receivers.

- Needle-nose or long-nose pliers, small size.
- Diagonal cutting pliers (dikes), small size.
- Pencil-type soldering iron, not over 30 watts.
- Pair of tweezers, about 6 inches long
- Dental probe—looking like a giant needle, with a flattened end for poking and scraping in tight quarters.
- Heat sink—a type of clamp which snaps shut in the manner of a clamp—to be used in absorbing any excess heat during soldering, thus preventing damage to parts.

One other tool, not exactly a must but very handy, is a type of soldering tool which automatically removes molten solder by suction, incident to unsoldering small parts, transistors, etc. During such unsoldering, very tiny mounting holes used for parts mounting, unintentionally fill up with solder, preventing the new part from being inserted; however, with a bit of experience and care, the dental probe in conjunction with the soldering iron can accomplish the same job. Also, the plugged hole can be carefully heated while a fine needle (not a wire!) is moved back and forth in the hole, thus dislodging the solder. Toothpicks work just as well if you can keep a supply on hand.

Fig. 10-1. Photograph of one type of sound module used in a Zenith TV.

PC WIRING

The printed-circuit technique eliminates old-fashioned point-to-point wiring by electrochemically printing the wires on a phenolic or fiberglass board. First, a layer of copper is chemically deposited on one or both (as required) sides of the board. Next, an etching process removes the copper from all areas except those to be used as conductors or connecting points. Fine holes are then drilled for inserting the fine wire leads of components to be mounted. Parts are then inserted in the proper holes and the board given a sold *bath* which solders all leads to their respective copper bands or areas. Figures 10-1 and 10-2 show examples of PC boards and mounted parts.

In removing a defective part heat is applied with a fine-tipped soldering tool to the immediate area of the joint. If feasible the heat sink clamp is attached close to this spot. Otherwise, a heat sink is improvised by pressing the flat end of a screwdriver blade or even a nail head, or by using the needle-nose pliers, so as to absorb the excess heat. The same precaution applies when soldering the new part into the circuit.

As in all soldering, and much more so here, speed is most important. The slower the soldering, the greater the risk of overheating and damage. With care and a little experience, a small joint can be made by a touch-and-go application of the iron. This method requires:

- A shiny, clean tip on the soldering tool
- A correspondingly clean spot on the work being soldered
- A good contact between the soldering tool and the work

A little dry-run practice on any piece of scrap material will soon make Mr. Do-it-yourself a proficient solderer.

A final note of good advice regarding the choice of solder. There is a great variety of solder on the market—from the coarse, lead-like stuff used by plumbers to the very fine, high-grade kind used in military equipment and other fine work. It is quite easy even for the novice to know which is which. Solder (especially the kind used in electrical work) is rated by the percentages of the mixture of lead and tin, the higher the proportion of tin, the better the solder; the shinier it also looks. Thus, a 40-60 grade is a relatively poor kind, having but 40% of tin. The 50-50 grade is commonly used in radio work, but is not recommended for fine work discussed here. The most suitable type is the 60-40 grade, having a 60% tin content. It is usually produced in a thin gauge, and flows much smoother than any of the others. A pound of this costs a few dollars, but quarter-pound will last the do-it-yourselfer a lifetime. Since this solder has a resin core (having a resinous chemical in its hollow core for cleaning surfaces during soldering) it is advisable to wipe the soldered area carefully and sparingly with alcohol to remove any resin residue, which might otherwise (in some cases) cause some electrical *leakage*.

EXAMPLES AND ILLUSTRATIONS

The following pages will illustrate the practical aspects of solid-state circuitry as far as the TV owner and the occasional do-it-yourselfer are concerned. These illustrations should be examined on a comparative basis, that is, in the light of your experience with vacuum tube and hybrid sets, based on the text and illustrations earlier in this book. This should enable you to make the transition to solid-state techniques with a minimum of difficulty. Illustrations and examples are not typical—they are actual portions of current TV receivers. As such, they are quite representative of modern TV industry practice. A note of clarification is in order: because of some inconsistency in terminology between different manufacturers,

Fig. 10-2. Another type of sound module used in a Zenith TV. The module shown in this photograph is not interchangeable with the one shown in the previous illustration.

Fig. 10-3. Module board layout of a modern solid-state TV (courtesy Magnavox Company).

Fig. 10-4. Photo of a modern GE solid-state TV receiver showing how to remove a module board for servicing. Test instrument shown on the left is a color-bar/crosshatch generator for converging the picture tube electron guns. Test instrument shown on the right of the receiver is a multimeter used for measuring volts, amperes, ohms, etc.

somewhat different names may be employed to denote the same basic unit. The term *module, PC board* and *circuit board* are all used at one time or another to denote what I have described as a printed-circuit board—a phenolic or fiberglass plate with electrochemically inlaid wiring and a number of parts or components preassembled thereto.

Figure 10-3 is an actual illustration of location of a number of circuit boards in a modern color receiver. This is but one of a number of illustrations found in the typical data package for the particular set in question.

Figure 10-4 illustrates the process of circuit board removal in one recent 1976-77 TV set. While exact locations and methods of removal and replacement will vary from set to set and among different manufacturers, this example illustrates current practice. As indicated before, the data for the specific set in question must of course be followed for exact procedures.

To repair or not to repair—There are a number of factors to be considered by the TV owner in deciding whether or not to tackle a PC board. First and most important is the capability of the individual. This includes previous experience, ease of working with small tools, as well as handling very small parts, fine soldering and general manual dexterity. Based on one's self-confidence with regard to the above, you may decide to go it alone or seek the services of the professional.

The second factor is the nature of the suspected or failed part. If, for example, an encapsulated submodule is determined to be defective, removal of the part and replacement is far preferable to the scrapping of a large PC board on which this submodule is mounted. On the other hand, when a relatively small circuit board on which the parts are soldered very flat against the board with practically no wires or leads showing is encountered, replacement of the whole unit is the wiser alternative. Finally, the cost of the replacement part is a factor. In considering the cost, the purchase price of the part as well as the repair shop bill should be taken into account, the latter almost always being the larger of the two.

To summarize, if after careful examination of the defective part or parts and an equally careful study of the applicable service information, you think that you can do the job, you will probably be successful. If, however, after all these preliminaries, you are afraid to tackle the job, the indication is that you are not up to the task, and the job should therefore be left to the professional TV servicer.

Chapter 11
Typical Trouble Symptoms and Their Probable Causes

In this chapter I have listed several common or typical trouble symptoms along with things that might cause such a malfunction. Bear in mind that these are *typical*. Being specific about specific troubles is a subject for an entire new book. These are intended to *guide* you in the right direction so that you can possibly pinpoint the exact trouble after checking over your TV.

Some of the headings indicating TV symptoms you have seen earlier in this book. Nonetheless they are presented here as an aid to refresh your memory after once reading this book. Just thumb through these pages to locate the trouble at hand. Then read through the listed possible causes, one by one, testing your receiver for defects.

When troubleshooting a TV receiver always keep a mental image of the block diagram in your mind. Or better yet, keep the actual diagram at your side. You'll find that by referring to the block diagram you can isolate the defective area much easier.

If you pull several tubes to test at your local supermarket or electronic parts house be sure to mark the location of each tube unless you have a tube layout diagram on hand. Even professional TV servicers have a difficult time remembering which tubes go where. Just outline the image of your particular TV chassis. Draw circles for the locations of the tubes with the tube type indicated inside each one. Also use care when handling the tubes. Besides being fragile the numbers rub off rather easily. If you can't tell what kind of tube it is it is of no use to you.

SET COMPLETELY DEAD

- Check AC outlet for power with a table lamp
- Check line cord and plug for possible defects
- Check line cord interlock at rear of TV
- Check circuit breaker or main fuse in TV
- See if dial lights or tubes light; if so check B+ fuse or circuit breaker
- Check on/off switch for proper operation

RASTER WITH NO SOUND OR VIDEO

- Check tuner tubes if applicable
- Check shielded cable from tuner to main chassis
- Clean tuner with spray lubricant
- Check IF amplifier tubes if applicable
- Check AGC pot setting
- Check noise gate pot setting if applicable
- Check video detector diode
- Check lead-in from back of receiver to tuner
- Check AGC tube if applicable

SOUND BUT NO RASTER

- Check for illuminated picture tube filament
- Check high-voltage rectifier tube if applicable
- Check high-voltage regulator tube if applicable
- Check damper tube if applicable
- Check horizontal amplifier tube if applicable
- Check horizontal oscillator tube if applicable
- Check horizontal circuitry fuse or circuit breaker if applicable
- Check video output tube if applicable
- Check for connection of high-voltage wire to the picture tube

RASTER & VIDEO BUT NO SOUND

- Check speaker by substitution
- Check audio output tube if applicable
- Check audio amplifier if applicable
- Check connection of wires between speaker and chassis
- Check connection of wires to and from volume control

- Check sound IF tubes if applicable
- Check sound discriminator/detector tube if applicable

LOUD ARCING OR POPPING NOISE

- Check connection of high-voltage lead to picture tube
- Check for defective rubber cup at picture tube where high-voltage lead connects
- Check lead inside high-voltage gauge
- Check neck of picture for blue arc (defective tube)
- Check ground strap around picture tube

NO RED & HUE CONTROL HAS NO EFFECT

- Defective 3.58 MHz crystal
- Check color oscillator tube if applicable

PICTURE IS REDUCED OVERALL

- Check low-voltage rectifiers (tube)
- Defective deflection yoke if raster has keystone appearance

HORIZONTAL LINE ONLY

- Check vertical oscillator and amplifier tubes if applicable
- Check vertical size and linearity controls for "dead" spots
- Check vertical leads to and from deflection yoke
- Check vertical hold control for damage

RASTER REDUCED TOP & BOTTOM

- Check vertical size and linearity adjustments
- Check vertical amplifier tube if applicable
- Check vertical centering control setting if applicable

RASTER APPEARS AS A SMALL CIRCLE

- Check position of deflection yoke on neck of picture tube

LACK OF WIDTH

- Check horizontal amplifier tube if applicable
- Check damper tube if applicable
- Adjust horizontal width device

TOO MUCH SNOW IN PICTURE

- Check tuner tubes if applicable
- Clean tuner with spray lubricant
- Check AGC pot setting
- Check noise gate pot setting if applicable

HORIZONTAL WEAVE WITH COLOR BARS

- Defective degaussing circuit—call a professional
- Defective low-voltage power supply—call a professional
- Shorted tube—check all tubes if applicable

MOVABLE GHOST WHEN FINE TUNING

- Check IF amplifier tubes if applicable
- IF alignment required—call a professional

VERTICAL ROLL

- Check vertical oscillator and amplifier tubes if applicable
- Check sync tube if applicable
- Check video amplifier and IF amplifier tubes if applicable
- Check AGC pot setting
- Clean tuner with spray lubricant
- Properly adjust vertical size and linearity controls

SMOKE OR FIRE

- Unplug receiver and call a professional

POOR B&W ON COLOR RECEIVER

- Check screen control settings
- Check drive control settings
- Check color amplifier tubes if applicable

PICTURE PULLING HORIZONTALLY AT TOP

- Check horizontal AFC tube if applicable
- Check AGC pot setting
- Check noise gate pot setting if applicable
- Check horizontal AFC diodes by substitution if applicable

BARBER POLING OR POOR COLOR SYNC

- Check color oscillator tube if applicable
- Check burst amplifier tube if applicable
- Check 3.58 MHz crystal by substitution
- Check color killer control setting

POOR FOCUS IN COLOR RECEIVER

- Check focus rectifier tube/diode by substitution
- Check focus control setting
- Check focus lead (heavy black wire) to picture tube at socket
- Check for excessive picture tube current by reducing brightness
- Check high-voltage rectifier tube/diode by substitution
- Check high-voltage regulator tube if applicable
- Focus divider assembly defective where applicable

MISCONVERGENCE

- Check for connection of convergence assembly plug on chassis
- Static convergence magnets displaced

REDUCTION IN SOUND AFTER WARMUP

- Check audio amplifier and output tubes if applicable
- Defective audio output transistor where applicable
- Dirty volume control
- Check sound IF amplifier tubes if applicable

IMPROPER TINT CONTROL RANGE

- Alignment problem—call a professional
- Check burst amplifier tube if applicable
- Check leads to and from tint control

SOUND BEAT PATTERN IN PICTURE

- Improper fine-tuning adjustment
- Improper operation of cable TV system where applicable
- Alignment problem—call a professional

AC HUM BAR FLOATING IN PICTURE

- Check *all* tubes if applicable
- Power supply or picture tube trouble—call a professional
- Antenna lead-in too close to power lines

ONE COLOR MISSING FROM PICTURE

- Check color amplifier tubes/transistors by substitution
- Check screen and drive control settings
- Check leads to and from picture tube base including socket.

Glossary

Glossary

AC—alternating current. The type of electricity normally used in homes and most industries. Its contrasting opposite is direct current, or DC, now obsolete except for certain specialized applications. All batteries supply DC.

ACC—automatic color control. A circuit similar in function and purpose to AGC, except that it is supplied exclusively to the color bandpass amplifiers to maintain constant signals.

AC hum—A low-pitch sound heard whenever AC power is converted into sound, intentionally or accidentally. The common AC hum is a 60-hertz note.

AFC—automatic frequency control. A method of maintaining the frequency or timing of an electrical signal in precise agreement with some standard. In FM receivers, AFC keeps the receiver tuned exactly to the desired station. In TV, horizontal AFC keeps the individual elements or particles of the picture information in precise register with the picture transmitted by the TV station.

AGC—automatic gain control. A system that automatically holds the level or strength of a signal (picture or sound) at a predetermined level, compensating for variations due to fading, etc.

amplifier—As applied to electronics, a magnifier. A simple tube or transistor or a complete assembly of tubes or transistors and other components may function as an amplifier of either electric voltage or current.

anode—The positive (+) element of a two-element device such as a vacuum tube or a semiconductor diode. In a television tube, an anode is an element having a relatively high positive voltage applied to it.

aperture mask—An opaque disk behind the faceplate of a color picture tube; it has a precise pattern of holes through which the electron beams are directed to the color dots on the screen.

arc—An electric spark that jumps (usually due to a defect) between two points in a circuit which are supposed to be insulated from each other, but not adequately so.

aspect ratio—The relation or proportion between the width and height of a transmitted TV scene. The standard aspect ratio is 4:3, meaning that the picture is 3 inches high for every 4 inches of width, or four-thirds as wide as it is high.

audio—Any sound (mechanical) or sound frequency (electrical) that is capable of being heard is considered as audio. Generally, this includes frequencies between about 20 and 20,000 hertz.

B+—Supply voltage, as low as 1 volt DC in transistor circuits and as high as hundreds of volts in circuits, which is essential to normal operation of these devices. The plus sign indicates the polarity.

B+ boost—A circuit in TV receivers which adds to, or boosts, the basic B+ voltage. The boost source is a by-product of the horizontal deflection system. Also see *damper*.

bandpass amplifier—In a color TV set, one or two color signal amplifiers located at the beginning of the color portion of the TV set; they are designed to amplify only the required color frequencies. They *pass* a certain *band* of frequencies.

Barkhausen—A term applied to a display of one or two black vertical lines on the left side of the picture tube due to some spurious behavior (oscillation) in the circuit. These lines are usually seen best when there is no picture on the screen (just a blank raster).

blanking—A term used to descibe the process which prevents certain lines and symbols, which are required for keeping the picture in step with the transmitter, from being seen on the TV screen.

brightness—Refers to both the amount of illumination on the screen (other than picture strength) and the control that is used to adjust the brightness level.

burst—In color TV, a precise timing signal. It is not continuous, but comes in spaced *bursts*. It is transmitted for controlling the 3.58 MHz oscillator essential for color reception.

burst oscillator—The precision 3.58 MHz oscillator vital to color reception. It is kept in step (sync) by the burst.

buzz—This is sometimes called *intercarrier buzz*, a raspy version of AC hum, usually caused by improper adjustment of some IF circuits.

B-Y—The blue component of a color picture minus the monochrome.

carrier—The radio signal which carries the sound or picture information from the transmitter to the receiver. The carrier frequency is the identifying frequency of the station (e.g., 880 kHz, 93.1 MHz, etc.).

cathode—The negative or minus element of a two-element tube or semiconductor. The cathode and the anode combine to form a diode (two-element device). The cathode is also the source of electrons in such devices.

cathode-ray tube—A tube in which electrical energy is converted to light. An electron beam (or beams), originating at the cathode, impinges upon a phosphor light-emitting screen. TV picture tubes, radar tubes, tuning eyes in some FM sets and many similar types are basically cathode-ray tubes.

cheater cord—An AC line cord for operating the TV set without the back cover or the cabinet when troubleshooting and repairing. The original cord is attached to the back of the cabinet as a safety measure.

chroma—Another term for color. Color amplifiers are often called chroma amplifiers. The term is also used to denote the control used to increase or reduce the color content of a picture.

clipper—A term describing the operation of one of the sync circuits in a TV set. It is the stage (tube or transistor) that separates the sync (timing) signals from the picture information.

color killer—A special circuit whose function is to turn off the color amplifier circuits when a black-and-white signal is being received. Also the control used to adjust the operation of the circuit.

contrast—The depth of difference between light and dark portions of a TV scene. Also the name given to the control for adjusting the contrast level.

convergence—The system that brings the three electron beams together in a color picture tube so they all pass through the same hole in the shadow mask and strike the correct dots on the screen.

converter—A stage in the tuner or front end of a TV set or any radio receiver which converts an incoming signal to a predetermined frequency called the IF or intermediate frequency. All incoming signals are converted to the same IF.

corona—Similar to an electric arc, except that this is a characteristic of much higher voltages (thousands). Corona occurs as a continuous, fine electrical path through air between two points, sometimes accompanied by a faint violet glow, usually near the picture tube.

crystal—A quartz of synthetic mineral-like slab or wafer having the property of vibrating at a precise rate or frequency. Each crystal is cut to vibrate at the desired frequency. Such a crystal is used in the 3.58 MHz oscillator to control its frequency.

damper—A diode, tube or semiconductor used in horizontal amplifier circuits to suppress certain electrical activity. It incidentally provides B+ boost voltage.

deflection—The orderly movement of the electron beam in a picture (cathode-ray) tube. Horizontal deflection pertains to the left-right movement, vertical deflection the up-down movement of the beam.

degaussing—Demagnetizing. In color TV sets, an internal or external circuit device which prevents or corrects any stray magnetization of the iron in the picture tube faceplate structure. Magnetization results in color distortion.

demodulator—A demodulator separates or extracts the desired signal, such as sound energy or picture information from its carrier.

detector—Same as *demodulator*.

diode—A two-element electron device, either a tube or semiconductor. The simplest and most common application of a diode is in the conversion of AC of DC (rectification).

discriminator—An audio detector in an FM receiver or TV sound circuits. Also, a detector performing a similar function in other frequency control circuits, such as horizontal frequency control.

faceplate—The front assembly of a picture tube. In a color tube it includes the tricolor phosphor and the aperture mask.

field—One scanning of the scene on the face of the picture tube, in which every alternate line is (temporary) left blank. The scan duration of a field is 1/60 second. Two fields, the second one filling in the blank lines left by the first one, make up a *frame*, or a complete picture. A frame duration is 1/30 second.

flyback, retrace—Name given to return movement of the electron beam in a picture tube after completing each line and each field. You don't see flyback or retrace lines (normally) on the picture tube because they are blanked out.

frame—The combination of two interlaced fields is called a frame. Since it consists of two fields each of 1/60 second duration, the frame duration is 1/30 second.

frequency—The number of recurring alternations in an electrical wave, such as AC, radio waves, etc. Frequency is specified by the number of alternations occurring during 1 second and given in hertz (cycles per second), kilohertz (1000 cycles) and megahertz (million cycles).

gain—Relative amplification. The number of times a signal increases in size (level) due to the action of one or more amplifiers. The overall gain of a signal often is millions of times.

gas—Refers to the presence (undesirable) of a trace of gas inside a vacuum tube. A gassy tube is a defective tube.

ghost—Most commonly a double-exposure type of a scene on the TV screen. Usually a fainter picture appears somewhat offset to the left of the main image, caused by the reception of two signals from the same station; one signal is delayed in time.

G-Y—The green color signal minus the monochrome.

high voltage—Generally refers to the multithousand picture tube voltage, but it can be used to mean any potential of a few hundred volts or more.

horizontal—Pertaining to any of the functions associated with left-to-right scanning in a picture tube including the horizontal amplifier, oscillator, frequency, drive, lock, AFC, etc.

hue—In color TV, the basic color characteristic which distinguishes red from green from blue, etc.

hum—Same as *AC hum*.

IC—module

IC—integrated circuit. A structure similar to a *module*, in which a number of parts required for the performance of a complete function are prewired and sealed. It is not repairable.

IF—intermediate frequency. In the tuner of a TV or radio receiver the incoming from the desired station is mixed with a locally generated signal to produce an intermediate signal, usually lower than the frequency of the incoming signal. The IF is the same for all stations. The tuner changes to accommodate each incoming signal.

in-line picture tubes—A more recent development in color tube structure which produces the three basic colors in adjacent strips or bars instead of the earlier types which produced three-dot or triad groups. Improved color quality as well as simplified design and maintenance is claimed for this type of design.

intercarrier—A term describing the current system of TV receiver design in which a common IF system is used both for picture and sound information. In older TV sets, the split-sound design was employed, in which separate IF channels for the picture and sound were used.

ion trap—See *trap*.

linearity—Picture symmetry. Horizontal linearity pertains to symmetry between the right and left sides of the picture, best observed with a standard test pattern. Also, an adjustment for achieving such linearity. Vertical linearity refers to symmetry between upper and lower halves of a picture.

line filter—A device sometimes employed between the AC wall outlet and a radio or TV set to reduce or eliminate electrical noises.

line, transmission—The antenna lead-in wire or cable.

lock, horizontal—An adjustment in some TV sets for setting the automatic frequency operation on the horizontal sweep oscillator.

loss—Usually refers to the amount of signal lost in the antenna lead-in (transmission line). This is particularly serious on UHF.

modulation—The process of combining (by superimposition) a sound or picture signal with a carrier signal for purposes of efficient transmission through air. The carrier's only function is to piggyback the intelligence.

module—A subassembly of a number of parts, usually including transistors and diodes. It is encapsulated and not repairable. See *IC*.

oscillator—Generator of a signal, such as the 3.58 MHz color subcarrier signal, the RF oscillator in the tuner, the horizontal oscillator signal (15,750 hertz) and vertical oscillator signal (60 hertz).

PC board—printed-circuit board. A subassembly of various parts, not necessarily all for one and the same function, on a phenolic or fiberglass board on which the interconnections are printed on metal veins or paths. No conventional wiring, except external interconnections, is used.

parallel—A method of circuit component connection where all components involved connect to common points so that each component is independent of all other components. For example, all light bulbs in your house are connected in parallel.

phosphor—The coating on the interior of the *faceplate* of a picture tube, which gives off light when struck by an electron beam. The chemical composition of the phosphor determines the color of the light it will emit.

power supply—That portion of a piece of electronic equipment which provides operating voltages for its tubes, transistors, etc.

preamplifier—A high-gain amplifier used to build up a signal so it is strong enough to present to the normal level amplifiers, for example, an antenna preamplifier for fringe area reception.

pulse—A single signal of very short duration used for timing and sync purposes. Sync pulses are the best example of this type of signal. Pulses occur in precisely measured bursts.

purity, color—The display of the various true colors without any accidental or unwanted contamination of one color by any of the others. Color purity is largely dependent on correct convergence adjustments.

raster—The illuminated picture tube screen fully scanned with or without video.

retrace—The return movement of the scanning electron beam from the extreme right to the extreme left and from the bottom to the top of the raster. Also see *flyback*.

retrace blanking—The extinction or darkening of the light on the face of the picture tube during retrace time in order to make these lines invisible. Should retrace blanking fail white lines sloping downward from right to left would be seen on the screen.

R-Y—sync separator

R-Y—The red color component of the overall color picture signal minus the monochrome.

saturation—Pertains to the full depth of a color, in contrast to a faint, feeble color. Saturated colors are strong colors.

semiconductor—A general name given to transistors, diodes and similar devices in differentiation from vacuum tube devices.

series—A connection between a number of components or tubes in chain fashion; i.e., one component follows the other. If any one component opens or burns out it breaks the series circuit.

shadow mask—Same as *aperture mask*.

shield—A metallic enclosure or container surrounding a component, tube, cable, etc. Also see *tube shield*.

shielded cable—A wire having a metal casing on the outside to prevent unwanted electrical energy from reaching the inner conductor.

signal—Electrical energy containing intelligence such as speech, music, pictures, etc.

signal-to-noise ratio—A mathematic expression which indicates the relative strength of a signal within its noise environment. A good signal has a high signal-to-noise ratio.

solid-state—A term indicating that the radio, TV set, etc., uses semiconductors and not vacuum tubes, but transistors, diodes, etc.

sound bars—Thick horizontal lines or bars, usually alternately dark and light, appearing on the TV picture screen due to unwanted sound energy reaching the picture tube. In appearance, the width, number, and position of these bars varies with the nature of the sound. Sound bars are caused by a misadjusted circuit.

subcarrier—The color picture information carrier. It is called a subcarrier because it is a secondary carrier in the particular channel. The color subcarrier frequency is 3.58 MHz.

sync—An abbreviation for a synchronizing signal. It is a timing signal or series of pulses sent by the transmitter and used by the receiver to stay in precise step with the transmitter.

sync clipper—See *clipper*.

sync separator—A circuit in a TV receiver which separates the sync from the picture information or the vertical sync pulses from the horizontal sync pulses.

trap—An electrical circuit which absorbs or contains a particular electrical signal (also called wave trap). Also a magnet, called an ion trap, used on the neck of some picture tubes for electron beam deflection.

triad—The three-color, three-dot group (red, green and blue) of which the color picture tube phosphor is made. Each group of three dots is a triad, and there are thousands of triads on a modern color tube screen.

tube shield—A metal sleeve which fits snugly over a glass tube and shields it from extraneous electrical impulses. A tube shield is part of the tube's electrical circuit.

UHF—ultra high frequencies. Radio and TV frequencies from 300 MHz upward. Channels 14 through 83 are all located in the UHF band and are, therefore, called UHF stations.

varactor—A semiconductor device having the characteristics of a tuneable device, through the application of a voltage. In contrast to the conventional frequency variation through the use of coil and capacitor techniques, the varactor requires only a voltage variation to effect tuning. In some recent TV sets varactor tuners have been used to replace the conventional coil-switching type tuners. Simplicity, greater stability and freedom from deterioration are claimed for this type of tuner.

vertical—Pertaining to the circuits and functions associated with the up-down motion or deflection of the electron beam.

vertical amplifier—An amplifier following the vertical oscillator used to enlarge the vertical sweep signal.

VHF—very-high frequencies. Radio and TV stations located below 300 MHz (down to 50 MHz). TV channels 2 through 13 as well as the FM band are in the VHF spectrum.

video—A term applied to picture signals or information (video, circuits, video amplifier, etc.).

VIR—vertical interval reference. Sometimes called broadcast controlled color correction system. It is an automatic, station-controlled signal used to initiate color and luminance corrective action in the TV set if so equipped. It is more elaborate and more effective than the automatic color control. See *ACC*. It is found in the more expensive TV receivers.

wave—Z-demodulator

wave—The name given to each recurring variation in alternating electric energy, including radio and TV signals.

X-demodulator—This is designation of the red (R-Y) signal demodulator.

yoke—Deflection yoke. The electrical assembly, somewhat in the shape of a yoke or collar, mounted on the picture tube neck against the flaring bell of the tube. By electric and magnetic means the yoke imparts to the electron beam the scanning (left to right and top to bottom) or deflection to produce the raster and the image.

Y-signal—This is the picture-only (minus color) signal that is fed to the color picture tube. It is sometimes called the brightness signal, meaning the actual brightness and darkness (and all shades in between) of the picture. This brightness signal plus the red signal produce all the red hues in the picture; the same Y-signal and blue give all the blue coloration and finally the Y-signal plus the green give the green coloration to the scene.

Z-demodulator—Same as the X-demodulator, except for the blue (B-Y) signal.

Index

Index

A

Adjustments, missing	165
AF amplifiers	15
AGC adjustment	121
Air waves	15
Alignment	31
Amplifier	14
AF	15
audio	35, 47
burst	68
chroma	67, 72
color	70
final	53
IF	32, 47
RF	15
vertical	36
video	34
Amplitude modulation	12
Antennas	13, 16
color TV	188
combination VHF-UHF	190
final	53
system	25
Antireflection device	65
Audio	14
amplifier	35, 47
frequencies	14
speaker	35
Automatic	22
brightness control	22
brightness limiter	22
Avoidance of distortion	95

B

Barkhausen interference	158
Bars,	
heavy black & white	151
sound	155
Blanking	20
Blooming	127
Blooming, when set is cold	130
Blue missing	184-186
Brightness control	149
Burst	57, 187
amplifiers	68
information	57
Buzz	130-136

C

Cage, high-voltage	39
Camera tubes	52
Carrier	22, 54
Centering	113
Centering controls	43
Cheater cord	81
Chroma	67, 187
amplifiers	67, 72
control	179
gain	169
Coax line	140
Color	61
adjustments	179-182
amplifiers	70
control	169
dots	61

functions	178
interdependence	182
killer	62, 71
level	169
presentation unstable	186-188
purifiers	191
purity	117
tube improvements, recent	63
weak	176
Consumer repair, advantages	194
Contrast control	120, 149
Control	187
Convergence	72
dynamic	72
static	62, 72

D

Deflection structure	40
Defocusing	127
Degaussing	191
Demodulator	16
Demodulators, color	70
Detail, loss of fine	161
Detector	16
sound IF	35
video	34, 54
Difference color system	179
Dim picture	148
when set is cold	130
Drive	179
Drive adjustments	76
Dynamic convergence	72

E

Electric signals	18
Electrical	
impulses	22
impulses into radio waves, conversion	22
Electron	18
beam	18
gun	40
stream	18

F

Field	20
Filaments	81
parallel tube	81
series-parallel tube	85
series tube	82
Fire	165-167
Frame	20
Frequencies	14
audio	14
intermediate	32
radio	14
Frequency modulation	12

Front ends	26
Functions	178
blocks	54
color	178

G

Gray-scale adjustments	75
Green missing	184-186

H

Halftone	18
Height	95
Herringbone weave	156
Horizontal	38
AFC	38
defects	117-120
deflection system	38
drive	43, 111
hold	43, 121
linearity	108
lock	122
lock-in	43
range	122
size	100
width	42, 100-108
Hum	130-133

I

IF amplifiers	32, 47
Image	17
dots	18
dots into electric signals, conversion	18
physical	17
Integrated circuits	47
Integrated circuits, applications	50
Interlacing	20
Intermediate frequencies	32
Ion trap	115

K

Kilohertz	15

L

Line	
noise	138
noise filters	138

M

Magnet precautions	116
Microphone	14
Mixer-oscillator	30
Modulation	12
amplitude	12
frequency	12
Modulators	52

N

No color but black & white ok	169-176
No sound but picture ok	143

O

Optical energy	18
Original	14
picture image to radio signal	17
sound to reproduced sound	14
Oscillator	57
color	57, 68
control	68
vertical	36
Overloading	120

P

Pairing	21
Pairing, line	163
Phosphor	60
Photographic recording	18
Picture	11
ghosts	159
image	17
information	54
multiple	136-138
snaking	154
system	11
system, AM part	12
tube circuits, color	74
tube, color	60
Pincushion correction	112
Points	106
Positioning	40
Pot	104
Pots, AGC	120
Power supply	44
Purifiers, color	191

R

Radio	14
frequencies	14
signal	17
waves	22
Raster only	142
Red missing	183-184
RF amplifiers	15
Receiver	16
actual	178
color	53
completely dead	141
solid-state, comparison	47
solid-state TV	46
tube-type, comparison	47
TV	26
typical	178
Receiving systems	13
Recording, photographic	18

S

Schematic diagram	80
Screen	40, 179
Sceen adjustments	76
Service techniques	195
Shadow mask	63
Short	165-167
Slug	104
Smear	163
Smear, color	191
Snow	144
Speaker	17
Soldering tool	80
Solid-state circuitry	197
examples	197-200
illustrations	197-200
Sound	
bars	155
IF detector	35
IF system	35
information	54
ok but picture missing	146
system	11
system, FM part	11
Static	62
convergence	62, 72
convergence, fixed	74
Strips	21
Sync	20, 181
circuits	36
color	68
failure, other casues	99
pulses	20
trouble, freak	100
Synchronizing	20

T

Terminology	12
Testing by substitution	92
Transistor	28, 80
Transmission line	26
Transmitter, color	51
Transmitter-receiver systems	11
Transmitting systems	13
Trinitron	63
Tubes	86, 87
all-glass	88-92
change, AGC	121
gassy	122
leaky	122
light only	142
octal	87
pulling	86
shield handling	88
with top caps	92

Tuners	26	Varicap	32
dual-purpose type	29	Vertical	36
noisy	32	amplifier	36
semiconductor	30	linearity	98
tube-type	26	oscillator	36
UHF	26	roll	98
varactor	30	size	95
VHF	26	slot structure	64
TV	11	sweep system	37
design, effect on	193	Video	34
receiver	26	amplifier	34
receiver, solid-state	46	detector	34, 54
set, hybrid	45	Visual spectrum, nature	51
signal, complete	24		
sound	23		
system	11		

V

		W	
V-output	96	Wiring	44
Varactor tuner, advantages	31	parallel tube	44
variable capacitor	32	PC	196
		series tube	45